Introduction
to
Rotary Drilling

by Annes McCann

Published by
Petroleum Extension Service
Division of Continuing Education
The University of Texas at Austin
Austin, Texas
1982

Catalog No. 2.01110
ISBN 0-88698-143-3

CONTENTS

Preface v

Lesson 1: Introduction to Rotary Drilling 1
    Background 3
    Rotary Drilling 3
    Petroleum Geology 5
    Geological Structures 7
    Application of Geological Concepts 9
    Petroleum Reservoirs 12
    Questions 17

Lesson 2: Rotary Drilling Rigs 25
    Introduction 27
    Derrick, Mast, and Substructure 29
    Power and Power Transmission 31
    Drawworks 32
    Blocks and Drilling Line 33
    Rotary, Kelly, and Swivel 34
    Circulating System 34
    Blowout Preventers 35
    Auxiliaries 36
    Rig Design Considerations 37
    Questions 55

Lesson 3: Mud Circulation Equipment 63
    Introduction 65
    Mud Pits 66
    Mud Mixers and Agitators 68
    Mud Pit Instruments 71
    Mud Storage and Handling 74
    Mud Cleaning Equipment 75
    Duplex Mud Pumps 82
    Triplex Mud Pumps 85
    Questions 89

Lesson 4: The Drill Stem 97
    Introduction and Early History 99
    Drill Pipe 99
    Tool Joints 110
    Drill Collars 114
    Drill Stem Auxiliaries 120
    Operations Involving the Drill Stem 122
    Questions 125

Lesson 5: Drilling Bits 135
    Introduction 137
    Roller Cone Bits 139
    Diamond Bits 153
    Drag Bits 156
    Questions 157

PREFACE

The material in this textbook is based on the IADC-PETEX Rotary Drilling Series and other appropriate training materials used by drilling contractors. The text is intended for industry and college-level petroleum technology students. Therefore, we have presented the information on a more advanced level. Special thanks should be given to Ron Baker and Dick Donnelly, PETEX staff content consultants, and Ferell Moughon, Sedco training specialist, for their help in adapting these materials.

Introduction to Rotary Drilling is Segment I in a three-part Drilling Technology Series. The new books are designed to respond to a need for more explanation of drilling procedures and of the math used in connection with those procedures. After reading the text, students can fill out the questions and hand in the pages to the classroom teacher for grading. When the lessons are returned, the student can insert them into his or her binder in order to have a personal manual for future study. (All three books are also available from PETEX as correspondence courses for those students wishing to take the course through the University of Texas.)

This segment of the series teaches background material necessary to someone planning for a career in drilling technology. The text covers basic petroleum geology and the rig and its equipment, and goes into some detail on the circulating system, the drill stem, and the bit. Segment II contains information necessary for conducting routine operations on a rig. Segment III teaches well control and other specialized practices. Some math calculations explained in the context of the appropriate rig procedures are presented in Segments II and III.

We hope this text and the other two segments will be of value to those students eager to have knowledge of drilling theory presented in the context of practical procedural instructions.

Annes McCann
Technical Writing Supervisor

Lesson 1
INTRODUCTION TO ROTARY DRILLING

Background

Rotary Drilling

Petroleum Geology

Geological Structures

Application of Geological Concepts

Petroleum Reservoirs

Lesson 1
INTRODUCTION TO
ROTARY DRILLING

BACKGROUND

Thousands of years have passed since man first scratched the surface of the earth in search of food, water, and a supply of energy. Oilwells are now being drilled to depths of almost six miles in the continuing search for the lifeblood of the modern world, fossil fuels. The first oilwell in the U.S. was a 69-foot hole drilled by Edwin Drake in Pennsylvania in 1859. More than 20,000 wells have now been drilled offshore; ocean-floor completions have been made below 1,500 feet of water, and the capability exists to complete such wells in 4,500 feet of water or deeper. Rotary drilling rig power has increased from 1 horsepower (hp) a hundred years ago to the 10,000-hp equipment now used offshore.

ROTARY DRILLING

The essential functions of rotary drilling rigs are *hoisting, rotating,* and *circulating.* The rigs consist of portable machinery and structures that can be quickly dismantled, moved, and reassembled on a new location. Rigs can be mounted on wheels or built on barges and ships to facilitate the transportation of equipment from rig site to rig site.

Rotary drill pipe is special upset-end pipe with thread and shoulder end connections (called *tool joints*). The pipe is carefully designed to withstand the tensile, torsional, and burst stresses of rotary drilling. Drill collars are heavy-walled — *Functions of Drill Collars* steel tubes placed at the bottom of the drill string to provide weight for the bit and hold the drill string in tension.

Rotary drill bits may be roller cone, diamond, or drag types. *Roller cone bits* may be of milled-tooth construction or have tungsten carbide inserts for teeth. They may have plain, lubricated, or journal-type bearings. *Diamond bits* have various configurations, but generally they feature fairly large stones held in a matrix that allows the diamonds to be pressed into the bottom of the hole when weight is applied. High-pressure fluid streams (jets) are used on *roller cone* and *drag* bits to facilitate the drilling process developed by applied weight and rotation.

The *hydraulic* system of a drilling rig provides a stream of — *Functions of drilling fluid* high-velocity fluid to clean the bottom of the hole and thus to improve the drilling process. The fluid also sweeps the cuttings out of the well and up to the surface. Drilling fluid may be made of water-base or oil-base mud, water, air, or gas. Water-base muds are the most common drilling fluids; they are made up of water, clay, inert solids, and chemicals in order to obtain the desired properties of viscosity, gel strength, and density.

The actual drilling process is influenced by the kinds of rock encountered, the rig capacity, and the types of drilling fluids and bits chosen. Hydrostatic pressure (the pressure of the drilling fluid at rest) has a significant influence on bit performance. The difference between hydrostatic pressure and formation pressure (the pressure of the fluids in the rock encountered) greatly influences the drilling rate. Other variables are weight and rotary speed applied to the bit.

Straight-hole drilling is actually a misnomer, since most so-called straight holes are usually within 2° to 3° off vertical. They do not change direction abruptly, and they do not have sharp edges or bends in the line of the hole from top to bottom. The pendulum effect of the drill collars at the bottom of the drill stem tends to make the bit drill a nearly vertical hole that has few sharp bends. Large-diameter drill collars and properly placed reamers and stabilizers make it possible to drill a straighter hole than an assembly of small-diameter drill collars without stabilizers or centralizers.

Directional drilling uses the principles of straight-hole drilling to form a wellbore that changes its angle in a desired direction according to a predetermined plan. Various types of *deflection tools* are used to establish the direction and amount of angle away from vertical. Downhole measurements and survey instruments are used to determine the direction and angle of a drilled hole.

Fishing is a drilling operation that involves recovering small equipment, drill pipe, drill collars, or whole strings of pipe that may be lost or stuck in an oilwell. The operation requires various catch tools, fishing string accessories, and wireline devices to survey and separate frozen strings of pipe.

Blowout prevention involves the entire hydraulic system of a rotary rig to control formation pressure. Preventing oilwell blowouts requires recognizing the preliminary signs of a blowout, utilizing the blowout preventers (BOPs), and circulating fluid of the required density to contain the formation pressure.

Well logging is used to evaluate oil or gas zones in a well. Electric logs can make estimates of what kind and how much production will occur. Open-hole logs can give readings of lithology, porosity, and hydrocarbon content. Acoustic and radioactivity logs make accurate assessments of porosity and estimations of permeability. Cased-hole logs can obtain much of the same information through the casing.

Casing and *cementing* are the first operations required to complete the well after the hole has been drilled, logged, and tested. The purpose of casing is to prevent the wall of the hole from caving in and to provide a means of extracting petroleum if the well is produced. Several strings of casing (heavy pipe)

Advantages of Drill collars —

Purpose of Casing —

4

are set before a well reaches final depth, but the *production string* is the casing set from the underground reservoir to the surface. The production string may be casing set from total depth to the surface, or it may be a liner set at some intermediate depth. Cementing is an operation that prevents — Benefits of cementing fluid flow between the formation and the casing. The cement is pumped around the casing to seal the annulus, to protect the casing from corrosion, and to prevent pollution of freshwater formations near the surface.

Well testing involves an early evaluation of the productive capacity of a well; it is usually performed in open hole before casing is set. Well testing may involve wireline formation test tools, drill stem testing, or formation evaluation based on core samples, electric log data, or other methods of downhole sampling. *Well completion* is the process of setting casing and providing a passageway for fluids to flow to the surface. Completing a well may involve setting a screen liner or perforating the production pipe. Running and setting a packer and tubing may also be involved in completing a well. Deep wells with extremely high pressures usually require special equipment to handle completion tasks.

PETROLEUM GEOLOGY

Geology is the science that concerns the history and life of the earth, especially as recorded in rocks. Since petroleum is an accumulation of past life buried beneath thousands of feet of rock, geological studies play an important part in finding oil and gas.

Early History of the Earth

The earth is thought to have originated some four to five billion years ago out of a condensing cloud of cosmic dust. Most theorists agree that the earth passed through a molten phase. During this phase, the components of the earth separated to produce a heavy core 4,400 miles in diameter, which was covered by a crust of the lightest materials some 10 to 30 miles thick (fig. 1.1). At about the same time, large amounts of water vapor and gases erupted to form the primeval atmosphere. The crust cooled slowly. All the rocks were of *igneous* origin, having been solidified from a molten form called *magma*. Masses of lighter rock were carried about on the surface of moving plates of heavier rock, becoming buckled and deformed as they collided or separated or slid past each other.

Rain began to fall on the cooling and developing atmosphere. The water ran from the heights and collected in the valleys and shallow depressions of the young earth to form the first oceans. Particles of rock were moved by the rain to lower places, where they finally settled out of quiet water as sediments. These sediments changed the face of the earth.

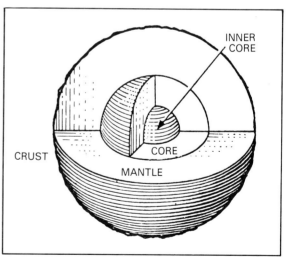

Figure 1.1. Crust, mantle, core, and inner core of the earth

5

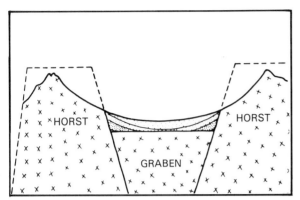

Figure 1.2. First sediments

Uplifted blocks of igneous rocks were worn down from their original form (fig. 1.2). This topographical change is shown by the broken line. Valleys between the uplifts were filled by the resulting sediment.

The earth was barren and lifeless for millions of years. Then at some unknown and ancient time, microscopic life began in the oceans. As such life forms died, they were buried in the sediments that continued to accumulate. The fossil remains of ancient animals and plants succeed one another in a definite and determinable order. Studying this order makes it possible to divide prehistoric times into eras and smaller subdivisions, the more important of which are shown in table 1.1. The number of years each subdivision lasted has been determined from radioactive studies of the minerals.

TABLE 1.1
GEOLOGIC TIME SCALE

Era	Period	Epoch	Duration (in millions of years)	Approximate Time (in millions of years ago)
Cenozoic	Quaternary	Recent	0.01	0.01
		Pleistocene	1	1
	Tertiary	Pliocene	10	11
		Miocene	14	25
		Oligocene	15	40
		Eocene	20	60
		Paleocene	10	70 ± 2
Mesozoic	Cretaceous		65	135 ± 5
	Jurassic		30	165 ± 10
	Triassic		35	200 ± 20
Paleozoic	Permian		35	235 ± 30
	Pennsylvanian		30	265 ± 35
	Mississippian		35	300 ± 40
	Devonian		50	350 ± 40
	Silurian		40	380 ± 40
	Ordovician		70	460 ± 40
	Cambrian		90	550 ± 50
Precambrian			4,500 ±	

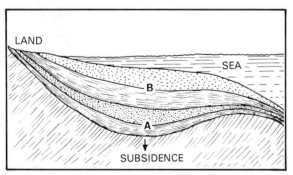

Figure 1.3. The weight of overlying unconsolidated sediments (B) compacts sediment layer (A) into sedimentary rocks.

The Rock Cycle

Erosion is the wearing away of rock or soil. It is caused by flowing water or the action of wind, freezing water, and waves. The original particles that eroded from the earth's surface were all derived from igneous rocks. Sediments continued to be deposited on top of other sediments (fig. 1.3). The earlier consolidated deposits (A) were compacted by the weight of the overlying rocks (B) and in the process were transformed into sedimentary rocks. These rocks in turn were eroded again to produce sediments, and thus the cycle of erosion and sedimentation continued.

GEOLOGICAL STRUCTURES

Formations

If some other process did not compensate for erosion, the land would by now have been reduced to plains near the level of the sea. Yet today the land stands about as high above sea level as it ever did. Obviously, uplift of the surface must have occurred to compensate for the wearing down of the mountains.

The earth's surface and upper crust have moved in all directions many times since the earth was formed. In fact, movements of the surface are continuing today, as can be seen by earthquakes that have recently occurred. Sometimes only a few feet of displacement can be seen along a fault after an earthquake. Geological evidence shows that repeated movements of only inches of the earth's surface may raise or lower thousands of feet of earth.

Sedimentary rocks are deposited in essentially horizontal layers, which are called *strata*, or *beds*. Table 1.2 classifies the components of sedimentary rocks.

TABLE 1.2
CLASSIFICATION OF SEDIMENTARY ROCKS

| CLASTIC | CHEMICAL | | ORGANIC | OTHER |
	Carbonate	Evaporite		
Conglomerate	Limestone	Gypsum	Peat	Chert
Sandstone	Dolomite	Anhydrite	Coal	
Siltstone		Salt	Diatomite	
Shale		Potash	Limestone	

Most rock layers are not strong enough to withstand the forces combating them, and they become deformed in some manner. One common deformation is the buckling of the layers into folds (fig. 1.4). *Folds* are the most common structure in all mountain chains, ranging in size from small wrinkles to arches and troughs that are miles in width. The upfolds or arches are called *anticlines,* while the downfolds are called *synclines* (fig. 1.5).

Anticlines and synclines both plunge. A short anticline with its crest plunging in all directions from a high point is called a *dome.* Domes often have a core that uplifts them, such as the salt domes along the U. S. Gulf Coast.

Earth Movement

Most rocks are fractured during an earth movement to form *joints,* which are actually cracks in the rocks. If the rock layers on one side of a fracture have moved with respect to the other side, the result is a *fault.* The displacement of a fault may range from a few inches to thousands of miles. The San Andreas fault in California is an extremely long fault line.

Figure 1.4. Schematic cross section shows deformation of earth's crust by buckling of layers into folds.

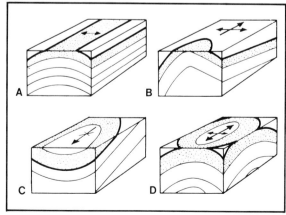

Figure 1.5. Simple kinds of folds: *A,* symmetrical anticline; *B,* plunging asymmetrical anticline; *C,* plunging syncline; *D,* dome with deep salt core.

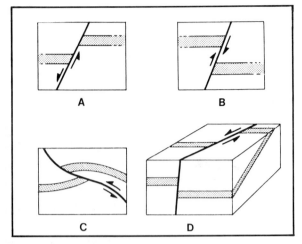

Figure 1.6. Simple kinds of faults: *A*, normal; *B*, reverse; *C*, thrust; *D*, lateral.

Faults may be classified into four major types—*normal, reverse, thrust,* and *lateral* (fig. 1.6). The names are derived from the movement of adjacent blocks. Movement is up or down in normal and reverse faults but horizontal in both thrust and lateral faults. A combination of vertical and horizontal movements is possible in all faults.

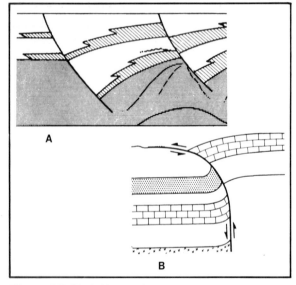

Figure 1.7. Variations of normal and reverse faulting: *A*, rotational faults; *B*, upthrust faults.

Normal and reverse faulting have variations (fig. 1.7). Such rotational faults and upthrusts have important effects upon the location of petroleum accumulations.

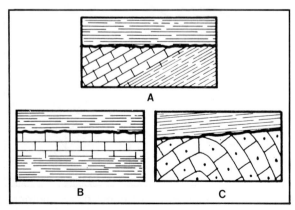

Figure 1.8. Two kinds of unconformities: *B*, disconformity; *A*, and *C*, angular unconformities.

Earth movements often erase or prevent the depositing of sediments that are present elsewhere. Such buried, eroded surfaces are called *unconformities*. One type of unconformity, a *disconformity* (fig. 1.8), is a break or abrupt change in the conformity or makeup of a formation. The beds above and below the surface are parallel. *Angular* unconformities are those in which the beds above the unconformity pass over the eroded edges of the folded beds below.

APPLICATION OF GEOLOGICAL CONCEPTS

The most successful method in the early days of oil exploration was to drill in the vicinity of *oil seeps,* places where oil was actually present on the surface. In fact, many of the great oil fields of the world were discovered because of the presence of oil seeps.

Oil seeps are located either up-dip or along fractures. In the upper diagram in figure 1.9, a seep at the outcrop of a reservoir bed is shown. Such seeps may be active where oil or gas is still flowing out slowly. An example is Mene Grande, Venezuela, only a mile up-dip from a large oil field. In other cases, the sands near the surface are completely sealed and the seep is no longer active, as at Coalinga Field, California. The Athabaska "tar sands" in Canada appear to be a seep of the Cretaceous age that was buried by later sediments and has now been exposed again by erosion. Seepage from fractures and faults (lower diagram, fig. 1.9) is common. The seep may be made up of oil, gas, or mud. Examples are the mud volcanoes of Trinidad and Russia.

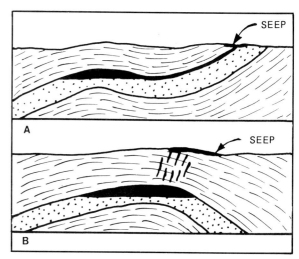

Figure 1.9. Location of seeps: *A,* up-dip; *B,* along fractures.

The presence of oil seeps on anticlinal crests was observed as early as 1842. It was not until after the drilling of the famous Drake well in Pennsylvania in 1859, however, that it was noted that newly discovered wells were being located on anticlines. Little practical use was made of this information until I.C. White applied it in search of gas in Pennsylvania and nearby states in 1885. During the latter part of the nineteenth century, geologists searched for oil in the East Indies and Mexico. In 1897, geological departments were established by some U.S oil companies. Many of the great discoveries made subsequently in the Mid-Continent, Gulf Coast, California, and other areas were made by applying geological knowledge.

Exploration Geophysics

By 1920, anticlinal folding was only one of a number of geological factors that were used to help predict oil and gas accumulations. Surface mapping alone left much to be desired. Fortunately, geophysical methods of exploration came into existence about this time. The torsion balance and the seismograph made it possible to predict subsurface structures. The seismic method introduced the transit time of sound waves generated by an explosion as an oil-finding technique. The transit times depended on the nature of the rocks that were being penetrated. Today, other methods as well as explosions are used to create sound waves for seismic testing. Devices such as the Vibroseis can generate continuous low-frequency sound waves. Under favorable conditions, a certain geologic bed can be mapped quite accurately by reflected and

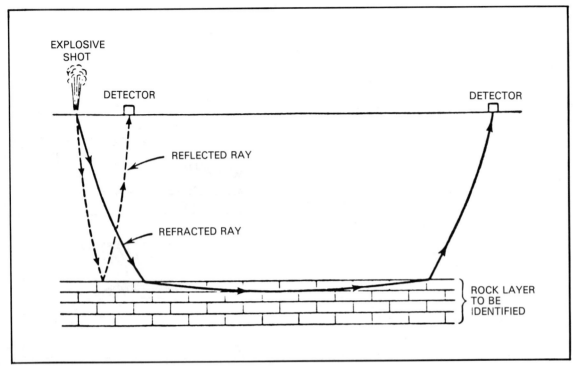

Figure 1.10. Reflected and refracted seismic waves

refracted seismic waves (fig. 1.10). The gravitometer and the magnetometer are other geophysical tools that make use of the physical properties of rocks to find structural conditions favorable to petroleum accumulation.

Subsurface Geology

The methods and techniques used in the study of subsurface geology have improved greatly since the 1920s. Today more oil and gas discoveries are credited to subsurface geology studies than to any other oil-finding method. Examination and correlation of cuttings, core samples, and wireline logs of various kinds yield important subsurface data. This information is used to prepare many kinds of maps and cross sections. Subsurface geologic structure can be shown by contour maps (fig. 1.11). Maps may show variations in the characteristics of the rocks and the structural arrangement, such as old shorelines or pinch-outs.

Of course, maps give only one view. To supplement the maps, vertical cross sections are made. These vertical cross sections may be designed to show structure or a particular detail of only a small interval. For example, the section in figure 1.12 shows the effect of a pinch-out of the darker colored sand body on the thickness between correlation markers *A* and *B* above and below it.

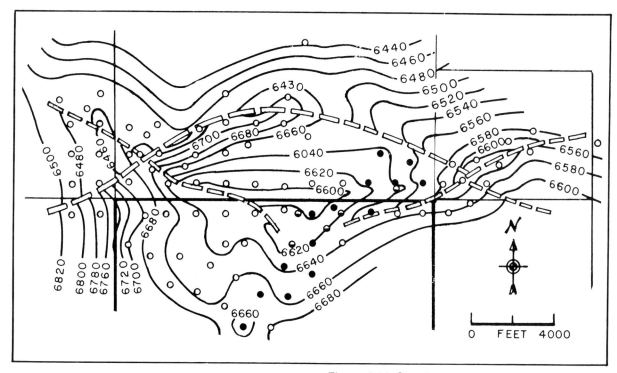

Figure 1.11. Structure contour map

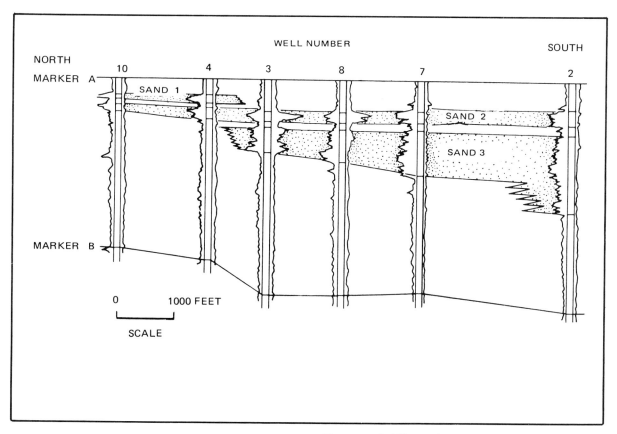

Figure 1.12. Effect of pinchout of sand 3

PETROLEUM RESERVOIRS

A petroleum reservoir must have (1) a source of oil and gas, (2) the existence of a porous and permeable bed, and (3) a trap that acts as a barrier to fluid flow so that accumulation can occur.

Origin of Petroleum

According to the *organic theory*, oil and gas originate from organic matter in sedimentary rocks. In the absence of oxygen, dead vegetation stops decomposing and accumulates in the soil as humus and as deposits of peat in bogs and swamps. Peat buried beneath a cover of clays and sands becomes compacted. As the weight and the pressure of the cover increase, water and gases are driven off. The residue, ever richer in carbon, becomes coal.

In the sea a similar process takes place. A rich variety of marine life is eternally falling in a slow rain to the bottom of the sea. Vast quantities are eaten or oxidized, but a portion of the microscopic animal and plant life is buried beneath the ooze and mud of the sea floor. This organic debris collects in sunken areas under a growing buildup of sands, clays, and more debris, until the sediment is thousands of feet thick. As the sediment builds, the pressure of deep burial begins to work. The extreme weight and pressure compacts the clays into hard shales. Within this deep unwitnessed realm of immense pressure and high temperatures, oil is formed.

Temperature is an important factor in the forming of hydrocarbons. The process apparently does not take place at temperatures less than about 150°F. Hydrocarbon generation is most efficient within the range of 225° to 350°F. Increasing temperatures convert the heavy hydrocarbons to lighter ones and ultimately to gas. However, at temperatures above 500°F the organic material is carbonized and destroyed as a source material. It is known that the earth's temperature increases as depth increases. Consequently, if source beds become too deeply buried by movements of the earth, no hydrocarbons will be produced.

Migration

Next in the process, the scattered hydrocarbons in the fine-grained source rocks are concentrated into a reservoir. Hydrocarbons tend to migrate upwards. Compaction of the source beds by the weight of the overlying rock provides the driving force necessary to expel the hydrocarbons and to move them up through the porous beds or fractures to regions of lower pressure. These regions of lower pressure are usually at a shallower depth. Gravity separation of gas, oil, and water also takes place in reservoir rocks that have been water-saturated. Gas (the lightest) rises, oil is in the middle, and water (the heaviest) goes to the bottom. Petroleum is, consequently, forever trying to rise to the surface until it is trapped or escapes. Vertical migration via faults and fractures has led to many large oil accumulations.

Reservoir Rocks

A petroleum reservoir is rock containing gas and/or oil, usually in combination with water. To be commercially productive, a reservoir must have sufficient thickness and pore space to contain an appreciable volume of hydrocarbons, and it must yield the fluids at a satisfactory rate when penetrated by a well.

Sandstones and carbonates are the most common reservoir rocks. The degree of a rock's *porosity* (pore space in relation to solid rock) is an important characteristic. The porosity may be *primary,* such as the natural intergranular porosity of sandstone. Or it may be *secondary,* due to chemical or physical changes such as dolomitization, solution channels, or fracturing. Porosity may be lessened by compaction and cementation. The degree of porosity in a reservoir results from numerous natural processes.

Sandstone porosity is controlled primarily by the mixing of the various sizes of grains, and by the way the grains are packed together. Porosity is at a maximum when grains are round and of uniform size; porosity decreases as grains become more angular. Artificially mixed clean sand has measured porosities of about 43 percent for extremely well-sorted sands. This figure is almost irrespective of grain size. The porosity decreases to about 25 percent for poorly sorted medium- to coarse-grained sands, while the fine-grained sands may have over 30 percent porosity. Many sands have only 10–20 percent porosity.

The ease with which fluid moves through the interconnected pore spaces of a rock is called the *permeability* of the rock. In 1856 Henry d'Arcy, a French engineer, devised a means of measuring the relative permeability of porous rocks. For this reason, numerical expressions of permeability are measured in *darcies.* Most reservoir rocks have average permeabilities considerably less than 1 darcy. Therefore, the usual measurement is in millidarcies (md), or thousandths of a darcy. Permeability varies from 475 md for a highly porous, well-sorted, coarse-grained sand to about 5 md for a fine-grained sand.

Compaction by the weight of the overburden will squeeze the sand grains closer together and at greater depths may crush and fracture the grains. The result is small pores, lower porosity, and a drastic decrease in permeability. Thus, a sandstone reservoir that could produce petroleum at 10,000 feet might be much too impermeable to be of any economic value at 20,000 feet. Cementation, which fills part or all of pore space, also tends to increase with depth.

The porosity, permeability, and pore space distribution in carbonates are related to both the type of sediment and the changes that have taken place after deposition. These changes may have formed fluid channels, or fissures.

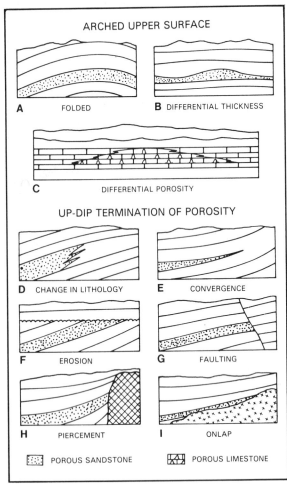

ARCHED UPPER SURFACE

A   FOLDED

B   DIFFERENTIAL THICKNESS

C   DIFFERENTIAL POROSITY

UP-DIP TERMINATION OF POROSITY

D   CHANGE IN LITHOLOGY

E   CONVERGENCE

F   EROSION

G   FAULTING

H   PIERCEMENT

I   ONLAP

PScSC POROUS SANDSTONE   POROUS LIMESTONE

Figure 1.13. Basic reservoir traps

Traps

Migration is a continuing process once the hydrocarbons have been expelled from the source rock. The hydrocarbons may move through the reservoir rock or through a fracture system. Obviously, a barrier or trap is needed to stop this migration so that the hydrocarbons will accumulate into an amount worth commercial attention.

A *trap* occurs when a set of geological conditions causes petroleum to be retained in a porous reservoir (or allows it to escape only at an extremely slow rate). For example, sandy shale can seal a porous and permeable sandstone reservoir formation.

Reservoir traps for oil and gas have two general forms. The trapping factor is either an arched upper surface or an up-dip termination of the reservoir. Some of the simpler forms are illustrated in figure 1.13. Diagram *A* shows an arched surface caused by folding, whereas diagram *B* shows an arched surface caused by different thicknesses of the sand bed. Diagram *C* represents a lenticular zone of porosity in an otherwise dense carbonate sequence where the arched upper surface is caused by differing porosity. (*Lenticular* refers to the lens shape of the porous zone.)

The lower six diagrams in figure 1.13 show the up-dip termination of the reservoir bed due to various conditions. Diagram *D*, for instance, shows change from sand to shale, whereas diagram *E* shows a sand bed pinching out because of lack of deposition. The middle diagrams show *truncation*, or cutting off, of the reservoir bed: diagram *F* shows erosion and trapping by the overlying beds above the unconformity; diagram *G* shows trapping at a fault by shale against sandstone. In diagram *H*, the trapping agent is the salt plug, and in diagram *I* the reservoir bed pinches out by onlap on an old land surface.

Reservoir traps for hydrocarbons are of two types: *structural* or *stratigraphic*, either alone or in combination. These traps have horizontal gas-water or oil-water contacts. Hydrodynamic (moving liquid) traps may also occur in different structural environments, but they are characterized by inclined gas-water or oil-water contacts.

Structural traps are traps that result from a deformation in the rock layer, such as an anticline or a fault. Although anticlinal traps vary widely in shape and size, they all have a common characteristic in that a gas-water or oil-water contact completely surrounds the accumulation of hydrocarbons. Fault traps depend upon the effectiveness of the seal at the fault. The seal may be the result of different types of formations being placed side by side (for example, shale against sand), or it may be caused by impermeable material called *gouge* within the fault zone itself. A simple fault trap may occur where structural contours provide closure against a single fault. However, in other structures such as a monocline, two or even three faults may be required to form a trap. In fault trap accumulations, the oil-water contact closes against the fault or faults and is not continuous, as in the case of anticlinal traps. Fault trap accumulations commonly tend to be elongated and parallel to the fault trend – for example, accumulations in the numerous oil fields along the Mexia-Talco fault zone, which extends from central to northeastern Texas.

The intrusion of underlying material, usually salt, into overlying strata often forms a variety of traps, both structural and stratigraphic. A *piercement dome* (fig. 1.14) may be circular, and is typical of the salt dome oil fields along the U. S. Gulf Coast and in Germany. Or it may be long and narrow as in the oil fields of Romania. The salt and associated material form an efficient up-dip seal. Hydrocarbon accumulations in the peripheral traps around a salt plug may not be continuous. Oil accumulations are usually broken into segments in smaller traps formed by modifying faults or structural closure against the plug. This discontinuous nature of oil accumulations in piercement domes is not favorable for development operations because it is unpredictable and thus increases the risk of dry holes.

Stratigraphic traps result from lateral change that prevents continued migration of hydrocarbons in a potential reservoir bed. Many are caused by the method of deposition, but others, particularly carbonates, are caused by later changes such as dolomitization, which tends to decrease porosity. Many large oil and gas fields are associated with this type of trap. The East Texas Field accumulation occurs in the truncated edge of the Woodbine Sand below an unconformity sealed by Austin Chalk (fig. 1.15).

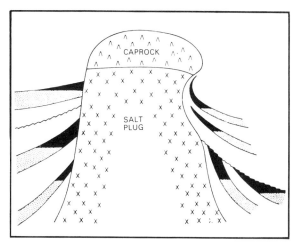

Figure 1.14. Piercement traps associated with a piercement salt dome

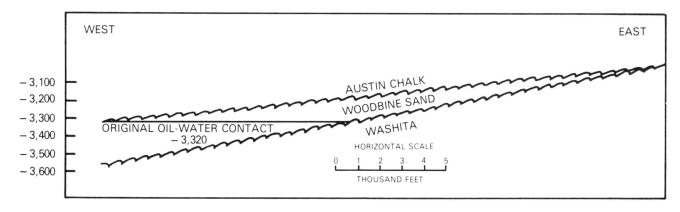

Figure 1.15. Pinchout trap between unconformities, East Texas field

Formation Pressures

In order to understand formation pressures (the pressure of the fluids in the rock pores), it is necessary to understand the concept of *hydrostatic pressure* (the pressure of a fluid at rest). Two factors affect hydrostatic pressure – the weight of the fluid and the depth at which it is tested. The deeper the measurement is taken, the greater the hydrostatic pressure becomes. The heavier the fluid is, the more its pressure will increase with depth. The rate at which the hydrostatic pressure of a fluid increases is called its *pressure gradient.* Heavier fluids have greater pressure gradients than lighter fluids. A pressure gradient for a specific fluid can be found out if the weight (or density) of the fluid is known, by consulting one of several published charts. Once the pressure gradient of a fluid is known, its hydrostatic pressure at any given depth can be figured; for hydrostatic pressure equals *true vertical depth* times a fluid's pressure gradient. (True vertical depth refers to the depth in feet of the point where the measurement is taken.)

Normal formation pressures are close in value to the hydrostatic pressure of a column of salt water. Along the U. S. Gulf Coast, normal formation pressure gains 0.465 psi (pressure gradient) per foot of depth. The geological explanation of normal formation pressure is that most porous rocks are connected to blanket-type sands, which have continuity through underground connections with porous formations that are exposed on the surface. Although the horizontal distance may be many miles, the true vertical distance is the depth of the formation. Pore pressure in porous and permeable formations, such as sandstones, is thus equivalent to hydrostatic pressure at the depth involved. By the same reasoning, pore pressure in a shale that is adjacent to a sandstone having normal pressure will be the same as hydrostatic pressure at that depth. Water will be squeezed out of the shale by the pressure of compaction until the pressure of the fluid in the shale equals the pressure of the fluid in the pore space of the sandstone.

Abnormal formation pressures develop in isolated reservoirs that are not connected through porous formations to the surface. As the result of compaction of the surrounding shales by the weight of the overburden of rock, water is expelled from the shale into zones of lower pressure, perhaps into a wholly confined sandstone that does not compact as much as the shale. Ultimately, a state of equilibrium will be reached, when no further water can be expelled into the sandstone, and its fluid pressure will then be approximately the same as the pressure in the pores of the shale. Actual pressure will depend on leakage to lower pressure zones, but in any case it will not exceed the weight of the overburden.

$$\text{Pressure gradient} = \frac{dP}{dz} = \rho g$$
$$= \text{weight density}$$
$$(N/m^3)$$

LESSON 1 QUESTIONS

Put the correct answer in the blank before each question. If there is more than one correct answer, put in all the correct letters. If a blank is drawn in the question, write out the answer as well as supply the letter in the multiple choice slot. The very act of writing down the answer will help you to remember it.

Look again at the **Background** and **Rotary Drilling Introduction** sections. These sections provide a basic introduction to the material to be covered in the *Drilling Technology Course*, Segments 1, 2, and 3. You will later go into much more detail in chapters devoted to these concepts. Answer now the following basic questions:

_____ 1. Drill collars are placed at the _____ of the drill string.
A. top
B. center
C. bottom ✓

_____ 2. The most common drilling fluid is _____.
A. gel
B. air
C. water-base mud ✓

_____ 3. In straight-hole drilling, the angle of the wellbore can be 10 to 15 degrees off vertical.
A. True
B. False ✓

_____ 4. A rig operation that involves retrieving items stuck or dropped in the borehole is called _____.
A. directional drilling
B. fishing ✓
C. logging

_____ 5. One way of preventing blowouts is to keep the _____ pressure in the proper relationship to the _____ pressure.
A. tubing; casing
B. hydrostatic; formation ✓
C. oil; gas

_____ 6. Wireline logging is used for evaluation in both open and cased holes.
A. True ✓
B. False

_____ 7. Purposes of casing pipe include –
✓A. providing a means of bringing the petroleum to the surface after drilling.
B. drilling the hole ever deeper.
✓C. keeping the wall of the hole from caving in during drilling.

_____ 8. Well completion may include –
    A. building the superstructure.
    B. completing the doghouse.
    ⌣C. perforation.
    ⌣D. setting a screen liner.

Look again at the **Petroleum Geology** and **Geological Structures** sections. Answer the following questions.

_____ 9. Studying the history of the deposition of rocks is important in predicting where accumulations of petroleum may be.
    ✓A. True
    B. False

_____ 10. The thickness of the earth's crust is _____.
    A. less than 5 miles
    ✓B. less than 50 miles
    C. more than 100 miles

_____ 11. Buckling of the crust is caused by –
    A. swelling in size.
    ⌣B. plate movement.
    C. organic decay.

_____ 12. Fossil remains show evidence of _____

    _____.
    A. twentieth-century rock deposits
    ⌣B. ancient plant and animal life
    C. the depths of the prehistoric seas

_____ 13. The earth is thought to have originated some 1 to 2 billion years ago.
    A. True
    ⌣B. False

_____ 14. It is impossible to estimate the length of geological eras.
    A. True
    ⌣B. False

_____ 15. The process of a wearing away of rock due to the action of water or wind is called

    _____.
    A. truncation
    B. deposition
    ✓C. erosion
    D. consolidation

_____ 16. The cycle of erosion and sedimentation causes the formation of

_____.
     A. igneous rock
     B. metamorphic rock
   ✓ C. sedimentary rock
     D. magma

_____ 17. Once the earth was formed, its surface stopped all movement.
     A. True
  ✓ B. False

_____ 18. Sedimentary rock is deposited in vertical layers.
     A. True
  ✓ B. False

_____ 19. Which statements are true about a dome?
     A. It is a type of fault.
  ✓ B. It is often caused by an uplifting core beneath it.
  ✓ C. Its sides slope in all directions from a high point.

_____ 20. If the rock layers on one side of a fracture have moved relative to the other side,

the result is called a _____.
  ✓ A. fault
     B. dome
     C. syncline
     D. disconformity

_____ 21. All of the pictures below are examples of different kinds of faults except –

A   B   C   D ✓

_____ 22. When a zone of porosity exists in a reservoir trap under a dense carbonate arch,
it is called a lenticular zone because it is shaped like –
     A. a round ball.
  ✓ B. a curved glass lens.
     C. a thin bed.

_____ 23. An up-dip termination of porosity in a reservoir bed can be caused by –
   A. truncation.
   B. pinching out.
   ✓ C. both of the above.

_____ 24. A fault trap works because a different kind of formation shifts to block the originally porous formation.
   ✓ A. True
   B. False

_____ 25. Salt dome oil fields have been produced along the U.S. Gulf Coast.
   A. True
   B. False

_____ 26. Both structural and stratigraphic traps may be associated with salt domes.
   ✓ A. True
   B. False

_____ 27. The famous East Texas Field was the result of _____.
   A. a dome
   B. a thrust fault
   C. an unconformity

Look again at the sections, **Application of Geological Concepts** and **Petroleum Reservoirs**. Answer the following questions.

_____ 28. A seismograph works because –
   A. various kinds of rocks record radioactivity differently.
   B. various kinds of rocks are more magnetic than others.
   ✓ C. various kinds of rocks take different amounts of time for sound to travel through them.

_____ 29. Maps produced from subsurface geologic studies can show only horizontal structure patterns.
   ? ✓ A. True
   B. False

_____ 30. A contour map shows a different level with each line.
   ✓ A. True
   B. False

_____ 31. Cross sections show a horizontal slice of land.
   ✓ A. True
   B. False

20

_____ 32. A commercially attractive petroleum reservoir must have all of the following except–
A. a trap.
B. permeability and porosity.
✓C. a natural outlet to the surface.
D. a source of oil and gas.

_____ 33. According to the organic theory, which of the following are needed to produce hydrocarbons?
A. Heat
B. Pressure
C. Microscopic plant and animal debris
✓ D. All of the above

_____ 34. If not trapped, petroleum tends to migrate to ever deeper areas in the earth.
A. True
✓ B. False

_____ 35. Which facts are true about the migration of oil?
A. Oil tries to reach areas of higher pressure.
✓ B. Oil needs a porous rock.
C. Oil can travel through impermeable rock.

_____ 36. The usual vertical order for fluids in an underground reservoir from bottom to top is _____.
A. oil, gas, water
✓ B. water, oil, gas
C. gas, oil, water

_____ 37. A reservoir's pore space in relation to its solid rock is the rock characteristic called _____.
A. permeability
✓ B. porosity
C. density
D. magnetism

_____ 38. Dolomitization is an example of primary porosity.
A. True
✓ B. False

_____ 39. Porosity may be lessened by which of the following actions?
✓A. Cementation
B. Fracturing
✓C. Compaction

_____ 40. The ease with which fluid moves through the interconnected pore spaces is a rock characteristic called _____.
   A. compaction
   B. dolomitization
   ✓ C. permeability

_____ 41. Most reservoir rocks have average permeabilities higher than 1 darcy.
   A. True
   ✓ B. False

_____ 42. *Formation pressure* refers to _____
   _____.
   A. fluid in the tubing
   ✓ B. the pressure of the fluids in the rock pores
   C. the pressure of the drilling mud

_____ 43. The hydrostatic pressure of a fluid decreases the deeper it is measured.
   A. True
   ✓ B. False

_____ 44. A heavier fluid has a greater pressure gradient than a lighter fluid.
   ✓ A. True
   B. False

_____ 45. A directional well that is slant-drilled for 2,000 feet has a true vertical depth of 2,000 feet.
   A. True
   ✓ B. False   ∠ 2000 feet

_____ 46. Pressure gradients for fluids of various densities can be obtained from published charts.
   ✓ A. True
   B. False

_____ 47. *Normal formation pressure* is found in a reservoir that has no opening to the surface.
   A. True
   ✓ B. False

_____ 48. Hydrostatic pressure is determined by multiplying _____
   _____.
   A. the horizontal measurement of the reservoir by its depth
   ✓ B. true vertical depth by pressure gradient
   C. abnormal pressure by normal pressure

_____ 49. *Abnormal formation pressure* is different from normal formation pressure in that –
   A. the reservoir has no connected opening to the surface.
   B. the pressure is greater.
   C. part of the pressure is caused by the weight of the overburden.
   ✓ D. all of the above.

_____ 50. A reservoir is 10,000 feet deep in a normal formation pressure area. Assume the pressure gradient to be 0.465. A well is drilled to 6,000 feet. What is the formation pressure at the well bottom?
   ✓ A. 2,790 psi
   B. 4,650 psi
   C. 60,000 psi

Lesson 2
ROTARY DRILLING RIGS

Introduction

Derrick, Mast, and Substructure

Power and Power Transmission

Drawworks

Blocks and Drilling Line

Rotary, Kelly, and Swivel

Circulating System

Blowout Preventers

Auxiliaries

Rig Design Considerations

Lesson 2
ROTARY DRILLING RIG

INTRODUCTION

A rotary drilling rig is a portable factory that is built to make hole. The requirement of *portability* (being able to be moved from site to site) places a limitation on rig design, as to both weight and size of each component. The total weight of the rig is a factor for overland moves, but the weight and bulk of each *unit* of assembled equipment is even more important. Each unit assembly is limited in weight because of truck and highway limitations on gross weight. Rotary drilling rigs must be disassembled for a move so that weight limits are not exceeded by any component or subassembly.

The most efficient rotary rig design is not necessarily the most effective. Other factors have to be considered. Rotary rig design should –

(1) allow for rapid erection and take-down, and provide for packaging in as few pieces as practical;

(2) not require special cranes for assembly (rig-up) or disassembly (tear-down);

(3) enable drill pipe to be run into the hole or pulled out with minimum time wasted;

(4) provide the maximum amount of available power for the circulating fluid to the bit (good hydraulics) when drilling.

Many factors determine a rig's portability. Wheel-mounted rigs can be used for drilling to depths of 10,000 feet or more and for completion/workover service on 15,000-foot wells. These rigs have self-erecting, telescoping masts; and the mast, drawworks, and engines are built on a trailer or self-propelled unit. Equipment such as mud pumps must be handled as packages; therefore, efficient planning and design are necessary.

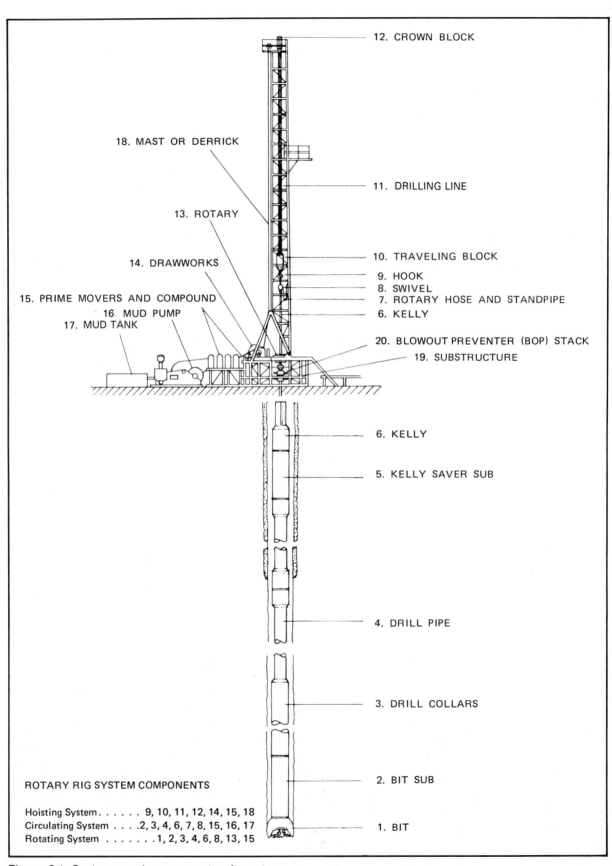

12. CROWN BLOCK

18. MAST OR DERRICK

11. DRILLING LINE

13. ROTARY

14. DRAWWORKS

10. TRAVELING BLOCK
9. HOOK
8. SWIVEL
7. ROTARY HOSE AND STANDPIPE
6. KELLY

15. PRIME MOVERS AND COMPOUND
16. MUD PUMP
17. MUD TANK

20. BLOWOUT PREVENTER (BOP) STACK
19. SUBSTRUCTURE

6. KELLY

5. KELLY SAVER SUB

4. DRILL PIPE

3. DRILL COLLARS

2. BIT SUB

1. BIT

ROTARY RIG SYSTEM COMPONENTS

Hoisting System 9, 10, 11, 12, 14, 15, 18
Circulating System 2, 3, 4, 6, 7, 8, 15, 16, 17
Rotating System 1, 2, 3, 4, 6, 8, 13, 15

Figure 2.1. Systems and components of a rotary
drilling rig

The relationships of the major systems of a land rig are shown in figure 2.1. These systems accomplish the three main functions of rotary drilling: *hoisting, rotating,* and *fluid circulation.* (Look closely at figure 2.1 as you read this introduction.)

The hoisting system is used to raise and lower the drill stem, and also to support and lower pipe that is used for casing and tubing. A mast or derrick supports the hook by means of the traveling block, wire rope, crown block, and drawworks. The drawworks is powered by two or three engines (called *prime movers*) to raise or lower the drill stem so that the bit can drill. The drill stem is the whole assembly from the swivel to the bit, including the kelly, drill pipe, drill collars, and bit sub.

A principal feature of the rotating system is the rotary table, or *rotary.* The rotary table is powered by the prime movers to rotate the kelly, which is raised or lowered through the kelly drive bushing. The rotation of the kelly causes the drill stem and bit to turn and thus "make hole" as the bit grinds away the rock formation. The kelly is supported by the hoisting system. Drilling fluid is pumped down the drill pipe to the bit and then up the *annulus,* the area in the hole that is outside the drill pipe.

The circulating system sends drilling fluid from a mud pit through the mud pump, standpipe, rotary hose, swivel, kelly, drill pipe, drill collars, bit, annulus, and back to the pit. The hydraulic power of the drilling fluid passing through the bit cleans the bottom of the hole and produces more effective drilling. Under special circumstances, a mud motor or turbodrill is used to turn the bit. In this case, hydraulic power of the drilling fluid (instead of rotation of the drill stem) turns the bit.

If rigs did not require mobility and quick rig-up and teardown capability, they could be designed to require less power for hoisting, pumping, and other jobs. Hydraulic rigs have been built, but they are heavy, slow, and troublesome to operate. The best means of hoisting drill pipe is the block-and-tackle arrangement that is generally employed.

DERRICK, MAST, AND SUBSTRUCTURE

Standard drilling rig *derricks* are tall steel structures with four supporting legs standing on a square base. A derrick is assembled piece by piece at the drilling site. A drilling *mast,* which is partially assembled when it is manufactured, usually has a smaller floor area. It is raised from a horizontal to a vertical position in one lift, as a beam supported at one end can be lifted. The standard derrick has become rare today except for extremely deep wells and offshore drilling. The mast (fig. 2.2) has almost completely replaced the conventional derrick for drilling on land because—

(1) it can be quickly dismantled and erected on another location by the regular rig crew; and

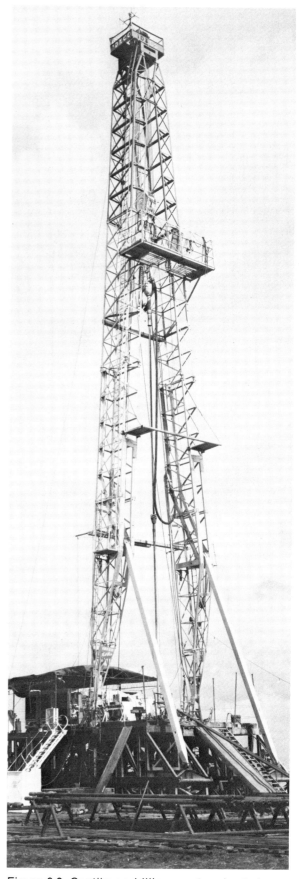

Figure 2.2. Cantilever drilling mast and substructure

29

(2) it can be moved in large units without complete disassembly.

Masts 135 to 145 feet in height are the most common size.

The mast or the derrick rests on a *substructure,* which also supports the rig floor and the rotary. The rig floor provides an area for handling the drill stem and related equipment. Blowout preventers and wellhead fittings are located under the substructure. The substructure supports the weight of the casing as it is being run in the hole. Drill pipe is suspended from the rotary table, which is supported by the beams of the substructure. Heavy-duty masts and substructures may have capacities of 1,200,000 pounds. The normal capacity is in excess of 500,000 pounds.

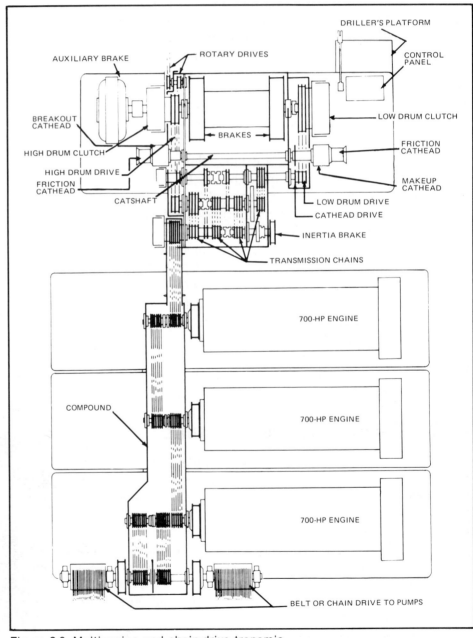

Figure 2.3. Multiengine and chain-drive transmission for a mechanical rig

POWER AND POWER TRANSMISSION

Steam is no longer a source of rig power, since natural gas (which was used to fire the boilers) has increased dramatically in cost. Most drilling rigs are now powered by internal-combustion engines and electricity. Large rigs and most wheel-mounted assemblies are generally powered by diesel engines. Diesel-electric rigs are fitted with diesel engines. These engines turn electric generators that supply AC or DC current to electric motors for the different rig machines.

Most prime movers are diesel engines, although some rigs are driven by engines that use natural gas or liquefied petroleum gas (LPG) in the form of propane or butane. Drilling rig engines range from 250 to 2,000 horsepower (hp) each; total rig power may be 500 to 5,000 hp. In figure 2.3, a conventional arrangement of the engines, compound, and drawworks for a mechanical-drive drilling rig is shown. In a diesel-electric system for rotary drilling (fig. 2.4), note the relative loads.

On a mechanical-drive rig, a means of transmitting the power from the engines to the drawworks, pumps, and rotary must be provided. This transmission is usually accomplished through an assembly known as the compound, which consists of clutches, couplings, shafts, chains, and sprockets.

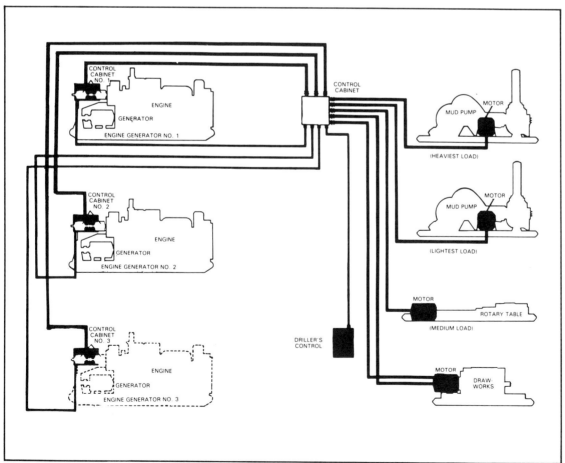

Figure 2.4. Diesel-electric system of power transmission

Figure 2.5. Drawworks for a heavy-duty rotary drilling rig

DRAWWORKS

A drawworks on a rig is known in other industries as a *hoist* (fig. 2.5). The main purpose of the drawworks is to lift and lower pipe in and out of the hole. The hoisting drum either reels in wire rope to pull the pipe from the hole or lets out wire rope to lower the traveling block and attached drill stem, casing, or tubing.

The drawworks includes a transmission, which uses chains, sprockets, and gears to allow speed changes of the hoisting drum. Often, the drawworks has a drive sprocket to power the rotary table. This arrangement is common, even on diesel-electric rigs.

The drawworks brake system makes it possible for the driller to control a load of several hundred tons of drill pipe or casing. Most rigs are equipped with two brake systems for the drawworks hoisting drum: one that is mechanical and one that is hydraulic or electric. The mechanical system consists of compounded levers to tighten brake bands to bring the drum to full stop. The hydraulic or electric brake can control the speed of descent of a loaded traveling block, although it is not capable of stopping the drum completely.

Another component of the drawworks is the catheads (fig. 2.6). The *makeup,* or *spinning, cathead* is located on the driller's side of the drawworks and is used to tighten the drill pipe joints. The other cathead, located opposite the driller's position, is the *breakout cathead.* It is used to loosen the drill pipe when it is pulled from the hole. Air hoists are provided on many rigs for handling light loads.

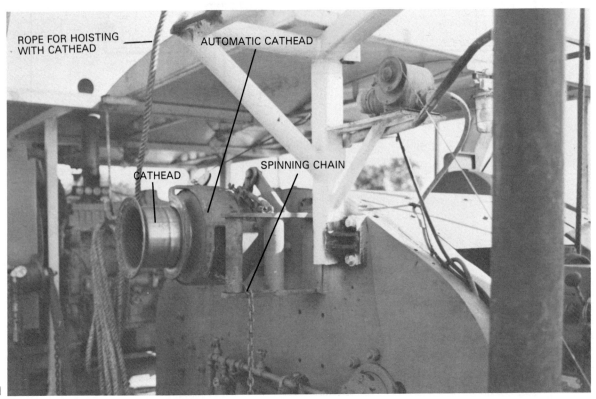
Figure 2.6. Cathead

BLOCKS AND DRILLING LINE

The *traveling block* (fig. 2.7), *crown block,* and *drilling line* within the derrick raise and lower loads of pipe out of and into the hole. During drilling operations, these loads usually consist of drill pipe and drill collars. The blocks and drilling line must also support casing while it is being run in the hole. This casing is often heavier than the drill stem. Drilling line is *reeved* around *sheaves* (pulleys) in the crown block at the top of the derrick or mast and in the traveling block (fig. 2.8).

The blocks and drilling line assembly must have great strength to support the heavy loads. The number of sheaves is determined by the weight to be supported. Five is the most common, but deeper wells often require six or seven. Friction is minimized in the blocks by heavy-duty bearings. Large-diameter sheaves are provided to lessen wear on the drilling line, which is usually a multistrand steel cable, 1¼ to 1½ inches in diameter. Wire rope requires lubrication to extend its useful life. The strands rub against one another as the rope flexes over sheaves in the traveling and crown blocks. Because wire rope eventually becomes too worn for use, it is an expensive, renewable item in the drilling process. A planned program of moving the wire rope allows longer service life. Ton-mile service records are maintained for slip-and-cut programs.

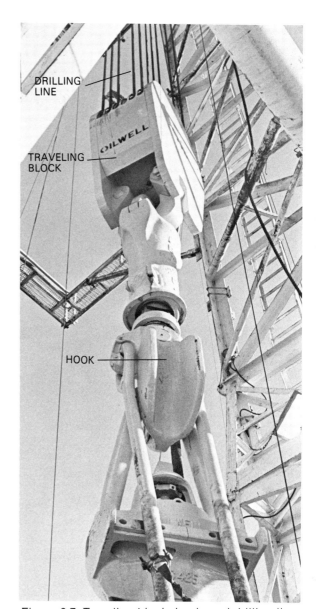

Figure 2.7. Traveling block, hook, and drilling line

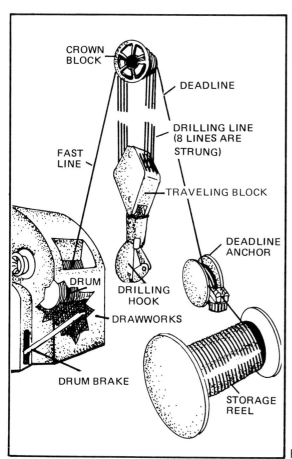

Figure 2.8. Hoisting system

Figure 2.9. The drill stem is turned by the drive bushing, which fits into the rotary.

Figure 2.10. Kelly, drive bushing, and rotary

ROTARY, KELLY, AND SWIVEL

The *rotary* is the piece of equipment that gives the rotary drilling rig its name; it is the machine that turns the drill stem and the bit in order to make hole. A rotary table is fitted with a drive bushing (fig. 2.9). The three-, four-, six-, or eight-sided kelly fits through the bushing and is thus turned by the rotary.

The rotary is a basic yet extremely rugged machine that is distinguished by its ability to withstand hard service. The drive bushing may fit in a square opening in the rotary table, or it may be driven by four pins that fit openings in the table. The drive bushing permits vertical movement of the kelly as the hole is deepened, at the same time rotating the drill stem. The rotary serves two main functions:

(1) to rotate the drill stem; and
(2) to hold friction-grip devices called *slips* to support the drill stem or casing.

The rotary may be mechanically driven by a sprocket and chain from the drawworks. However, many drilling rigs provide power to an electric motor that drives the rotary directly. In some cases, an independent engine is used to drive the rotary.

The *kelly* is the top member of the drill stem. It is about 40 feet long and may be either triangular-, square-, hexagon-, or octagon-shaped to fit its drive bushing. The kelly can move freely up and down through the drive bushing while it is being turned by the rotary. (fig. 2.10).

The *swivel* (fig. 2.11) hangs from a hook under the traveling block, and serves several vital functions.

(1) It supports the weight of the drill stem.
(2) It allows rotation of the drill stem.
(3) It provides a passageway for drilling fluid to enter the drill stem.

The *rotary hose* is connected to a gooseneck-fitting on the swivel; drilling fluid is pumped into the gooseneck, through the swivel, and down the kelly. This fluid may be under pressure exceeding 3,000 psi.

CIRCULATING SYSTEM

An essential feature of the rotary drilling process is the circulating system, commonly called the mud system. In order for rotary drilling to proceed, the drilled cuttings must be lifted out of the hole. Fluid must be pumped down through the drill stem and out the bit nozzles, and back up through the annulus (the space outside the drill string).

The principal purposes of circulating fluid are –
(1) cleaning the bottom of the hole by washing the cuttings back up to the surface;
(2) cooling the bit;
(3) supporting the walls of the well; and
(4) preventing entry of formation fluid into the borehole.
The circulation fluid is usually a liquid, but it may be air or gas. Water is the usual base, though occasionally oil is used.

A pump forces the drilling fluid up through a standpipe hose into the swivel, down through the drill stem, and back to the surface again (where it returns to the mud pits). The mud pits or tanks are usually fitted with solids-control equipment, which removes cuttings and other solid material in mud brought up from the hole before it is recirculated into the well by the mud pump.

When air is used as drilling fluid, the mud pump is replaced by compressors and there is no need for storage pits and settling tanks. Compressed air is forced down the drill stem to the bit and up the annulus by air pressure.

Figure 2.11. The swivel is attached to the upper end of the kelly.

BLOWOUT PREVENTERS

Drilling fluid in the hole helps prevent formation fluid from entering the borehole. If formation fluid does enter the well, it may rise to the surface and cause some of the drilling fluid to flow out of the hole. If the flow can be controlled by the drilling crew, it is called a *kick*. If the flow is continuous and cannot be controlled, a *blowout* has occurred.

A *blowout preventer* (BOP), in conjunction with other equipment and techniques, is used to shut off and control a kick before it becomes a blowout. Several BOPs are usually installed on top of a well (fig. 2.12), with an annular preventer above and two or more ram preventers below. An annular preventer has a resilient sealing element. When activated by fluid pressure, the sealing element closes on the kelly, drill pipe, or drill collars. Ram preventers have two steel ram segments that are pushed together from both sides to seal around drill pipe. Both annular and ram preventers are operated by hydraulic fluid pressure. Blind ram preventers can be used to close an open hole (hole with no drill pipe in it).

Blowout preventers are opened and closed by hydraulic power. The fluid is stored under pressure in an *accumulator*. High-pressure lines carry the hydraulic fluid from the accumulator to the BOP stack. When the driller turns the proper valves, the fluid operates the BOPs. Because the preventers must be able to close quickly, the hydraulic fluid is put under 1,500 to 3,000 psi of pressure by nitrogen gas in the accumulator unit.

Figure 2.12. Modern high-pressure blowout preventers

35

AUXILIARIES

Electric Generators

Modern rotary rigs provide power for auxiliaries with AC generators that are usually diesel-powered. Most of these generators have capacities of 50 to 100 kilowatts, although larger units are sometimes installed. The generators have enough capacity to carry the main power load of the rig (excluding hoisting, pumping, and rotating functions). A second engine and generator unit are held in ready reserve. AC electricity is used for rig lighting, shale shaker motors, mud pit stirrers, centrifugal pumps, rig instruments, engine-cooling fans, air conditioning for bunkhouses, and other purposes.

Air Compressors

A small compressor is usually mounted on the engine compound for supplying air to the pneumatic controls and clutches. The compressor has a volume tank to allow reserve storage of compressed air. Large rigs usually have another electrically powered compressor to furnish high-pressure air for other purposes, such as starting the main engines and operating air-powered hoists, air slips, BOP equipment, water wells, and air-operated tools.

Water Pumps

Water supply is an important item for drilling rig operations. Water is usually obtained from a well, stream, lake, or pipeline from a remote source. A stored supply of several hundred barrels is maintained at the rig. This may be in a pit or tank(s) of sufficient capacity to maintain operations for a short time if the primary supply is interrupted. Low-pressure water pumps are usually provided for washdown and for cooling the brakes of the drawworks. High-capacity pumps are generally used for mud and cement mixing and mud transfer.

Other Equipment

Drilling rigs also include such facilities as fuel storage tanks, a house for changing work clothes, a doghouse (a small structure on the rig floor that serves as an office for the driller), a place to store parts for the pumps and other equipment, and other facilities. Most large rigs are provided with an office trailer where the supervisors can maintain communications with the head office.

RIG DESIGN CONSIDERATIONS

Factors

When assembling equipment and machinery to make up a rotary drilling rig, the following factors that affect performance and mobility must be considered:

(1) range of well depths to be drilled;

(2) sizes of drill pipe and drill collars to be used for expected hole diameters;

(3) casing sizes and loads expected to be handled;

(4) size of table opening, rotary speed and power needed;

(5) circulating fluid volume, pressure, pump size and power required;

(6) mud system requirements, (i.e., mud pits, piping, pit accessories, dry mud storage, etc.);

(7) conventional derrick or mast and type of substructure that will be used (their rated capacities and heights);

(8) blowout preventers and control equipment necessary;

(9) auxiliary services required (i.e., electricity, compressed air, water supply); and

(10) miscellaneous items such as instruments, housing, and pipe racks needed for efficient operation.

Hole depth, drill pipe sizes, and expected sizes and weights of casing will determine the derrick load capacity. The size of casing will establish the rotary table opening. Rotary revolutions per minute (rpm) and drilling weights to be used will dictate rotary power and speed options. The volume and pressure of mud circulation will establish the size of the pump and its pressure rating. Deep, large-diameter holes will require more mud in the active mud system than shallow, slim holes. Conventional derricks require rig builders for assembly; and often a crane is required on the new location before the rig is transferred. Most drilling masts are cantilever types that can be assembled quickly by the rig crew and raised to vertical position by using the drawworks.

If high-pressure BOPs are required, wellhead clearance beneath the rig floor may exceed 20 feet. Manually operated blowout preventers may be used on shallow, low-pressure wells, but larger preventers with pressure ratings up to 5,000 psi are operated hydraulically.

Rig lighting, shale shakers, and mud mixers require electricity, usually 110- or 220-volt AC power. Miscellaneous electrical power totals as much as 100 kilowatts for heavy-duty rigs. Many rigs require compressed air for rig controls, brakes, air hoists, and other purposes. Washdown, mixing procedures, and brake cooling require readily available fresh water under pressure. Miscellaneous items of equipment that must be considered include instruments such as the weight indicator, pressure gauges, mud monitors, and rate-of-penetration indicators, usually placed in the doghouse on the rig floor. A crew change facility and quarters for supervisors and others are usually self-contained units.

High Substructures

High substructures pose two major problems:

(1) getting the drawworks on to the rig floor; and
(2) aligning extra-long chain drives from the engines to the drawworks.

These problems can be circumvented by step-down substructures for the engines.

Derrick or mast substructures with 20 to 30 feet of clearance for the wellhead and blowout preventers require cranes or precarious truck ramps for placing the drawworks on the rig floor. Substructure height may be reduced by using rig layouts with alternative designs. If the drawworks is on the derrick floor on a mechanical rig, the engines must be close to the same height to avoid long chain drives. Electric-drive rigs cut down on substructure size by eliminating engines on the derrick floor. They can be removed to ground level, since no alignment is necessary between the engines and the rig motors. When the drawworks is mounted at, or near, ground level, the substructure need be no larger than that needed for the derrick floor. Mast erection can also be simplified by hinging it near ground level.

A rotary rig with either a derrick or a mast can be assembled with hoist, transmission, and engines on the ground. A *catworks* with a drive arrangement for the rotary is placed on the rig floor. The drawworks brake lever is operated from the floor, using an extension linkage. Mechanical rigs that use this design, with a propeller shaft from the ground-mounted transmission to drive the catworks and rotary, are available. Ground-mounted hoist assemblies not only eliminate unnecessary substructures but also lighten the main substructure and speed up erection and dismantling.

Offshore Platforms

Since offshore drilling rigs are so expensive to install, they are often designed for the use of *multiple-well derricks*. A multiple-well derrick consists of a standard derrick that has been widened to provide space to drill two, four, six, or even more than twenty wells without the necessity of moving the derrick. All the holes are drilled from the same platform. Only the crown block, the racking platform, and the rotary table must be moved for the different wells located within the one derrick (fig. 2.13).

For drilling in deep water from a floating platform, rigs must be designed to withstand the side-to-side whipping action

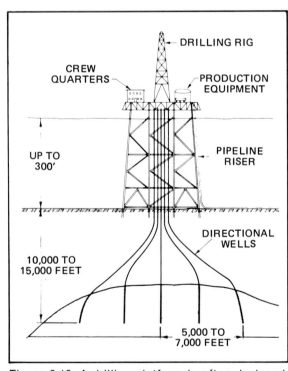

Figure 2.13. A drilling platform is often designed for directional drilling.

produced by the roll of the sea. Some method must be provided to prevent the traveling block from striking the side of the derrick during drilling under rough sea conditions. One means is a traveling block guide, which may be either a wireline strung from the crown to the derrick floor or a rigid track-and-roller arrangement.

Offshore derricks are also subject to loosening of bolts from sea action during drilling or trauma while mobile units are moved from place to place. Specially welded derricks have been designed to withstand such problems.

Rig Drive Systems

The major power trains in a modern rotary drilling rig are—

(1) drawworks and mechanical transmission from the engines or electric motors;
(2) wire rope system, including crown block, traveling block, wire rope for hoisting;
(3) mud pump drives;
(4) rotary drives; and
(5) power transmission devices.

Today the efficiencies of the first four components are greatly influenced by the type of power transmission devices chosen. Rigs are referred to according to their method of power transmission, as follows:

(1) mechanical drilling rig;
(2) DC/DC drilling rig;
(3) SCR electrical drilling rig.

The following paper presents a comparison of the various costs and efficiencies of the three different power transmission systems. Conclusions and opinions offered in the paper are, of course, the author's own.

THIS IS A PREPRINT — SUBJECT TO CORRECTION

AN ECONOMIC EVALUATION OF ELECTRIC DRILLING RIGS

By

Will L. McNair

Vice President-Engineering
Baylor Company

Abstract

Electric Drilling Rig Construction and usage have increased significantly over the past five years to replace both mechanical compound rigs and DC-DC systems. While it is generally conceded that the electric rig is more economical to operate, very little quantitative data and analysis has been developed to aid in this evaluation.

This paper presents a cost evaluation of electric rigs versus mechanical and DC rigs, including initial cost, operating cost, maintenance cost, repair cost, and lost revenue due to down-time. Suggestions for operating procedures to reduce cost are also presented. Understanding the electric rig operational advantages and properly training the drilling crew are significant areas where costs can be affected and these are outlined in the presentation. A typical example of rig cost comparison is presented to give the rig operator savings he should achieve with the electric rig.

Introduction

With continuing inflation of fuel and labor costs, the improved performance and flexibility of the electric drilling rig can result in operational savings over mechanical or DC-DC systems. Understanding the factors that contribute to costs on an electric rig will allow the operator to keep his costs to a minimum and his drilling rate to a maximum.

Among the most significant factors that should be considered are:

1) Initial equipment cost.
2) Proper utilization of diesel-generator sets to optimize fuel efficiency.
3) Keeping electrical losses to a minimum.
4) Proper training of operating personnel.
5) Maintaining an effective maintenance program and spare parts purchases.
6) Proper rig management and cost accounting to identify variances from budgeted areas.

Information herein describes the three basic power systems and their relative economic performance and provides guidelines to improve the operating effectiveness of an SCR electric drill rig. Supporting data, curves and calculations are provided in the appendix of the paper.

It should be emphasized that the estimates and assumptions included herein are used for a typical cost evaluation example of a drill rig. Actual analysis of a particular system and its data should be carried out, following the guidelines presented in the paper.

Summary

The SCR electric drill rig offers considerable flexibility for maximum availability of the rig for drilling. Load assignment functions allow one motor load to be transferred quickly from one SCR converter to another in the event of failure. Multiple engine-generator sets operate to a common AC bus which allows redundancy in event of a diesel or generator failure, ability to use fewer engines for maximum loading and efficiency, and provides power for auxiliary loads without an additional E-G set.

Improved fuel economy, lower maintenance, faster rig up/rig down time and improved availability are items in favor of the SCR type rig. A typical 12,000 foot SCR land rig shows estimated cost savings of $1.6 million over a similar mechanical rig and a cost savings of $700,000 over a DC-DC type rig for a 3 year operating period.

The most significant cost factor in operating a rig is the cost of diesel fuel. While the efficiency of electrical equipment can influence the fuel consumed, the most inefficient element in the power system is the diesel itself. Paying a premium for quality, economical, engines when buying the rig will pay off in fuel savings during operations.

A Review and Comparison of Rig Power Systems

The three basic rig power systems in use today are:

SCR drives, where several diesel-AC generator sets provide AC electrical power to a common bus. This power is further controlled by Silicon Controlled Rectifier (SCR) converters to power DC motors for variable speed drilling functions. This same AC electrical bus provides lighting and auxiliary function power without additional generating sets.

Mechanical drives, where diesel engines are coupled directly to equipment or through compound shafts to drive the rotary, drawworks and mud pumps. Separate diesel-AC generator sets are used for electric lighting and auxiliary functions.

DC-DC drives, or Ward-Leonard systems, where a diesel engine drives a DC generator to power DC motors for variable speed drilling functions. Separate diesel-AC generator sets are used for electric lighting and auxiliary functions.

To further understand the effectiveness of the SCR electric rig, an efficiency evaluation of each power system is shown below and described in Figures 1, 2, and 3.

SCR Drive Efficiency - SCR converters exhibit 98-99% efficiency, depending on loading, and AC generators are approximately 95% efficient. Considering 2-1/2% cable, DC motor blowers, and field losses, an SCR drive system efficiency from the engine shaft to the drilling machinery input is approximately 86-1/2%.

Mechanical Drive Efficiency - From the engine shaft output to the drilling machinery input, chains, shafts, and torque converters are used to transmit the power. Mechanical efficiency of these components vary but generally comprise an average drive efficiency between 70% to 80%, or 75% average.

DC-DC Drive Efficiency - Electrical generators and motors are approximately 95% efficient in the 800-1000 HP range. Considering further losses in the DC generator field power requirements (1%), blower cooling power (1%) and, cable losses (1/2%), the DC-DC drive efficiency from the engine shaft to the drilling machinery input is approximately 87-1/2%.

Even though the DC-DC system shows a 1% efficiency improvement over the SCR system, an additional engine-AC generator must be dedicated to the auxiliary AC power generator. When considering overall equipment utilization, the mechanical drive diesels usually run constantly regardless of their load and the DC-DC system must have an additional diesel-DC generator for the independent rotary. Also, when factoring in the diesel requirements, the SCR systems offer greater overall flexibility and efficiency when generator shedding at light loads is utilized.

To gain a better insight into the effect fuel consumption has on operating costs over a three year (25,000 hour) period, Table 1 compares several factors involved, including initial rig cost, fuel cost, maintenance cost and non-productive drill time for a typical land rig as described in Figures 1, 2, and 3. (All data estimated and calculated in Appendices A through G.)

TABLE 1
3-Year Period

	Mechanical	DC-DC	SCR
Total Rig Cost	$3,600,000	$3,600,000	$3,600,000
Total Rig Diesel Engine Power	2700 KW	2700 KW	2400 KW
Main Engine Costs	$250,000	$250,000	$225,000
Estimated Average Diesel Power Produced	1400 KW	1300 KW	1200 KW
Energy Consumed - KW-HRS	33 million	30 million	27 million
Estimated Average Diesel Efficiency of System	9 KW-hrs/gal	10 KW-hrs/gal	11 KW-hrs/gal
Calculated Fuel Consumed for 25,000 hours	3,670,000 gals	3,000,000 gals	2,450,000 gals
Diesel Fuel Costs for 25,000 hours @ $1.00/gallon	$3,670,000	$3,000,000	$2,450,000
Estimated Maintenance Cost Savings Over Mechanical (3 years)	$ --	$3,600	$7,200
Estimated Rig Up/Rig Down Savings Over Mechanical (3 years)	$ --	$126,000	$126,000
% Availability Savings Over Mechanical (3 years)	$ --	$165,000	$330,000

Summarizing Table 1

Summarizing these figures shows the following cost savings of an SCR electric unit over a 3 year period.

1. Savings compared to a mechanical rig:

	Savings
Initial Rig Cost	$ 0
Fuel Costs	1,222,000
Maintenance Costs	7,200
Rig Up/Rig Down Costs	126,000
Availability Income	330,000
Savings	$1,683,200

2. Savings compared to a DC-DC rig:

	Savings
Initial Rig Cost	$ 0
Fuel Costs	550,000
Maintenance Costs	3,600
Rig Up/Rig Down Costs	0
Availability Income	165,000
Savings	$718,600

Optimizing Electric Rig Performance

To effectively optimize the performance of SCR electric rigs, it is essential that the equipment functions be understood. The main features of the rig power system include:

** Multiple paralleled AC generators that can easily be added or removed from the system bus based on its load.

** Ability to quickly reassign a motor load from one SCR converter to another in the event of failure or maintenance.

** Adjustable rotary table torque limits.

** Engine load sensing provided to avoid engine overloads.

** The AC power generated for drilling, hoisting, and mud pumps is also utilized for lighting and auxiliary motors without additional engine-generator sets.

** No realignment of mechanical compound shaft after rig move.

The following checklist will aid the rig operator to improve his costs.

1) Make the decision to invest in fuel efficient diesels even though their initial investment is greater. Pay-back in fuel cost savings is fast.

2) Keep diesels operating near their full load rating to allow more efficient loading of engines.

3) Keep only those electrical auxiliaries running that are essential to rig operation; turn off all others.

4) Keep mechanical equipment well lubricated and maintained since mechanical losses consume fuel.

5) Purchase quality drilling bits, pipe, and drill mud to reduce the number of trips and fishing jobs which consume non-productive fuel that could be used for drilling.

6) Use "energy efficient" AC motors on auxiliary functions.

7) Apply AC motors near their nameplate rating to keep motor operating at high efficiency and high power factor.

8) Keep all electrical components clean and dust-free to prevent overheating and voltage breakdown.

9) Do not undersize interconnecting cables that might result in additional electrical losses.

10) Keep electrical power contactors and switch controls clean to reduce electrical losses and reduce failure rates.

Training Personnel

One of the most serious obstacles to effective operation is that of inadequately trained personnel to operate and maintain the rig equipment. The drilling industry has, historically, used mechanics for maintaining the equipment since mechanical drives were primarily used. These mechanics were also taught fundamental electricity in order to replace a defective motor or generator or to check wiring to a circuit breaker.

Technology is now such that greater knowledge in electronics repair and maintenance is desirable. While the supplier of the SCR equipment can aid significantly in repairing their equipment, it is advantageous for the rig operator to understand its operating principles and the basics of troubleshooting.

Each drilling contractor should train his personnel to understand the SCR type rig through sources such as:

1) Technical seminars

2) IADC sponsored training schools

3) Supplier sponsored schools

4) Films, articles, books

Not only should toolpushers and rig electricians attend, but superintendents, engineers, managers, and salesmen should become familiar with the SCR electric rig.

It is further suggested that the process of hiring new rig maintenance personnel should be reversed. Instead of hiring mechanics and then teaching them electrical fundamentals, hire electricians and teach them mechanical maintenance on the rig.

Establishing a Maintenance and Spare Parts Program

While the first step in effective rig operation is to understand its functions, the second step is to establish a disciplined maintenance program. The most effective method is to prepare a list of maintenance items and their frequency of inspection and replacement and then assigning responsibility for these inspections.

In general, the SCR electrical equipment requires little maintenance except for those devices that carry and interrupt heavy currents. Such items include DC contactors, circuit breakers and brushes/commutator on DC motors.

Keeping the electrical equipment clean from dust, oil, exhaust particles, and moisture will contribute significantly to continuous operation without failure.

When equipment is first purchased, a complete list of spares should be established to facilitate quick repair in the event of failure. It should also be established whether certain parts should be kept at the rig site, at the rig contractor's home office, or stocked at the supplier's facility. Consideration for domestic or foreign drill sites, as well as the probability of failure of certain items, will also affect a spares philosophy. Close coordination with equipment suppliers will aid in establishing a good spare parts program.

Specifying an Electric Rig

Once it has been established where the drill site will be, its local geology, and the depth of drilling expected, selection of equipment can be made. Selecting mud pumps, drawworks, mud system, and rotary table will be based on experience, availability of equipment and cost. Once these essential components are established,

the electrical power and control system can be defined. The sequence of defining the power system will follow the general guidelines below:

1) Perform a drilling profile study to determine the peak power and average power along the drilling route. Include the auxiliary and lighting power requirements.

2) Using the estimated efficiency of motors, SCR units and transformers, determine the peak and average kilowatts (KW) and KVA required from the 600 volt system bus.

3) Select the engine-generator set, rating, and quantity that will deliver the peak and average KW and KVA throughout the drilling cycle. Consider redundancy as required.

4) Prepare a list of all auxiliary functions, including motor horsepower, lighting KVA, etc. to establish the motor control center requirements.

5) Incorporate all of the above data into a specification that includes the following:

   a) Drawworks with electric auxiliary brake type and its size;

   b) Mud pump type and its size;

   c) Direct or independent rotary and its type and size;

   d) Auxiliary function list of equipment;

   e) Diesel engine type, size, quantity;

   f) AC generator type, size, quantity;

   g) SCR ratings and quantity required;

   h) Driller's Control functions required.

From this information, the supply store or equipment supplier can provide quotations and equipment to suit the contractor's needs at optimum cost.

Rig Cost Management

As with any business, effective cost accounting and controls should be maintained on drill rigs. Once an effective cost system is established, it can be used to identify variances from what is expected and to improve the rig operating efficiency.

Since fuel and labor costs are on the increase at significant rates, some creative experimentation on the rig should be tried to reduce these costs. Shutting down certain diesel-generator sets at appropriate times, turning off unused equipment, and similar activity, can be evaluated as to its cost effectiveness if a good cost accounting system is used.

Reference is made to the 1979 IADC paper presented by Richard C. Parsons of Santa Fe Drilling Company entitled "Drilling Rig Cost Accounting Techniques" as a good example of an effective cost accounting system.

Appendices

The items included in these appendices are calculations, curves and data used to support the previous analysis. These calculations can be used as a guide for rig operators to evaluate their particular rig.

Appendix A

Estimated Rig Up/Rig Down Time Evaluation

It is generally understood in the industry that a mechanical drive system must be properly aligned throughout in order for it to function properly after a rig move. Diesel-electric drives are mounted on a skid and alignment between rotating units is maintained with a minimum of adjustment. Experience shows that 1-2 days of rig up time are saved with an electric system.

If a particular land rig is moved every 90 days, the savings over a 3 year period would amount to 12-24 days. Assuming 18 days saved as an average with a 5 man set-up crew, the cost savings would be:

1) Set-up Labor Savings
   18 days × 5 men × $200/day = $18,000

2) Drilling Revenue Gained
   $6,000/day × 18 days   = $108,000

   Total Saved   = $126,000

3) Savings on a DC-DC rig over mechanical would be similar: $126,000

Appendix B

Estimated Maintenance Cost

Since the SCR system does not have the extensive power transmission of a mechanical rig, it requires less time to perform lubrication, chain repair, bearing replacements, etc., on the rig. A rough estimate of 1 man-day per month of maintenance, minimum, can be saved with the SCR system.

1) Over a 3 year period (36 months):
   36 man-days × $200/day = $7,200 savings.

2) A DC-DC system would have approximately 1/2 of savings, or $3,600.

Appendix C

% Availability

Because of the fast repair time associated with the SCR rig and the fact that load assignments can be changed in the event of a failed SCR power unit, the availability for drilling is greater than for a mechanical unit. The mechanical system would require down time for repair in the event a major failure occurred in the compound transmission. While actual figures on rig availability due to the power system are not readily available, it is estimated that a 5% improvement in rig availability is achievable with an SCR system.

If the rig day rate is $6,000, the savings over a 3 year period would be:

1) 3 year savings = 55 days × $6,000/day = $330,000.

2) Because a DC-DC system requires dedicated engine-generator sets for each function, some of its flexibility is lost in the event of a failure. Repair time would be increased, causing greater down-time before it would be fully operational. It is estimated that the DC-DC system availability savings would be 1/2 that of the SCR system, or $165,000.

Appendix D

Electric Rig System Power Analysis of Figure 1 For a 70 Day Drilling Cycle For a 12,000' Hole

I. Assumptions

Average Mud Pump 1 Power - 800 HP
Average Mud Pump 2 Power - 400 HP
Drawworks Power - 1500 HP Peak, 30%
    duty cycle
Average Rotary Power - 300 HP
Average Auxiliary AC Power - 350 KVA for
    drilling and tripping; 200 KVA for
    testing and casing
Drill Cycle:
    Drilling - 41 days
    Tripping - 6 days
    Testing, casing, cementing - 18 days
    Rig up, rig down, move - 5 days

II. SCR System

  A. Drilling Cycle:

    1) Mud Pump 1
       $800 \text{ HP} \times .746 = 597 \text{ KW}$
       Referred to 600 volt bus:
       Power = $597 \div (.95 \times .99) = 635$ KW
       KVA = KW $\div$ PF = $635 \div .7 =$
         907 KVA

    2) Mud Pump 2
       $400 \text{ HP} \times .746 = 298 \text{ KW}$
       Referred to 600 volt bus:
       Power = $298 \div (.95 \times .99) = 317$ KW
       KVA = KW $\div$ PF = $317 \div .7 =$
         453 KVA

    3) Rotary
       $300 \text{ HP} \times .746 = 224 \text{ KW}$
       Referred to 600 volt bus:
       Power = $224 \div (.95 \times .99) = 238$ KW
       KVA = KW $\div$ PF = $238 \div .6 =$
         397 KVA

    4) Auxiliaries
       350 KVA
       Referred to 600 volt bus:
       KVA = $350 \div .95 = 368$ KVA
       Power = KVA $\times$ PF = $368 \times .8 =$
         295 KW

    5) 600 volt bus total:
       Total Power: MP1 + MP2 + R + A =
         635 + 317 + 238 + 295 = 1485 KW
       Total KVA: 907 + 453 + 397 + 295 =
         2052 KVA
       Equivalent Power Factor: KW $\div$ KVA
         = $1485 \div 2052 = .72$

    6) For three balanced AC generators,
       each will deliver:
       Power = $1485 \div 3 = 495$ KW
       KVA = $2052 \div 3 = 684$ KVA
       Power Factor = .72

    7) Each diesel will transmit only real
       power through its shaft:
       Output Power = 495 KW $\div$ .95 EFF =
         521 KW
       Total output energy for 41 days per
         diesel = 521 KW $\times$ 41 days $\times$ 24
         hrs/day = 512,664 KW-Hrs
         (1,538,000 KW-Hrs for 3 engines)
       From diesel efficiency curve, the
       fuel consumed per engine:
       Gallons/diesel engine = 512,664
         KW-Hrs $\div$ 12 KW-Hrs/Gal =
         42,722 gallons for one engine
       Total fuel for 3 engines = 42,722 $\times$
         3 = 128,166 gallons

  B. Tripping Cycle:

    1) Drawworks
       $1500 \text{ HP} \times .746 = 1119 \text{ KW peak}$
       Referred to 600 volt bus:
       Peak Power = $1119 \div (.95 \times .99) =$
         1190 KW
       Peak KVA = $1190 \div .8$ PF = 1487 KVA

    2) Auxiliaries
       350 KVA
       Referred to 600 volt bus:
       KVA = $350 \div .95 = 368$ KVA
       Power = $368 \times .8$ PF = 294 KW

    3) 600 volt bus total
       Total Peak Power = DW + Aux. =
         1190 + 294 = 1484 KW
       Total Average Power = $1190 \times 30\%$
         + 294 = 651 KW
       Total Peak KVA = 1487 + 368 = 1855
         KVA
       Total Average KVA = $1487 \times 30\%$ +
         368 = 814 KVA

4) For three balanced AC generators, each will deliver:
Peak Power = 1484 ÷ 3 = 495 KW
Average Power = 651 ÷ 3 = 217 KW
Peak KVA = 1855 ÷ 3 = 618 KVA
Average KVA = 814 ÷ 3 = 271 KVA

5) Each diesel will transmit only real power through its shaft:
Output Peak Power = 495 ÷ .95 EFF = 521 KW
Output Average Power = 217 ÷ .95 = 228 KW
Total Energy Output per Engine = 228 KW × 6 days × 24 hrs = 32,832 KW-Hrs (98,446 KW-Hrs for 3 engines)
Estimated Diesel Efficiency = 6.5 KW-Hrs/Gal
Gallons/Diesel Engine = 32,832 ÷ 6.5 = 5,051 gallons
Gallons for 3 Engines = 5,051 × 3 = 15,153 gallons

C. Testing, Casing, Cementing Cycle:

1) Auxiliaries
200 KVA
Referred to 600 volt bus:
KVA = 200 KVA ÷ .95 = 210 KVA
Power = 210 × .8 = 168 KW

2) Each of 3 generators will deliver:
Power = 168 ÷ 3 = 56 KW
KVA = 210 ÷ 3 = 70 KVA
To improve efficiency, use only one engine-generator during this cycle; turn all others off.
Generator will deliver 168 KW and 210 KVA

3) Engine will deliver: 210 ÷ .95 = 221 KW
Energy Output = 221 KW × 18 days × 24 hrs = 95,472 KW-Hrs
Fuel economy will be approximately 6 KW-Hrs/Gal
Gallons used = 95,472 ÷ 6 = 15,912 Gals

D. Summary of Calculations:

1) Total KW-Hrs consumed in power and drilling equipment for 70 day drill program:

Drilling	1,538,000
Tripping	98,496
Testing, cementing, casing	95,472
Total KW-Hrs	1,731,968

2) Total fuel consumed in 70 day drill program:

Drilling	128,166
Tripping	15,153
Testing, cementing, casing	15,912
Total Gals	159,231

3) Fuel costs for drilling and finishing hole @ $1.00/gal in 70 day drill program: 159,231 × $1.00 = $159,231

4) Cost per KW-Hr: $159,231 ÷ 1,731,968 = $0.09 (compared to commercial utility costs of $0.04 per KW-Hr)

5) Equivalent diesel engine efficiency: 1,731,968 KW-Hrs ÷ 159,231 gals = 10.88 KW-Hrs/Gal

6) Projected fuel costs for one year: 365 ÷ 70 × $159,231 = $830,276

7) Additional data obtained from this study is:
Maximum KVA required from each generator = 835 KVA

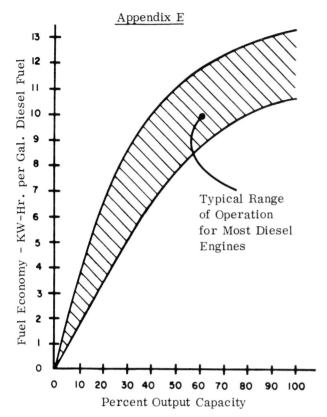

Appendix E

Typical Range
of Operation
for Most Diesel
Engines

Typical Diesel Engine Efficiency Curve
Showing Effect of Loading and Range of
Economy

(Note: Range of operation includes 2 and 4
stroke engines, natural and forced injections
units, at sea level.)

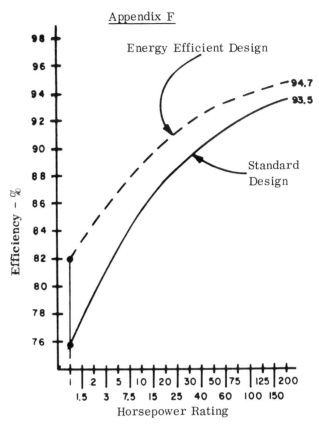

Appendix F

Energy Efficient Design

Standard
Design

Industry Average of Efficiency Versus
Horsepower for 4-Pole, 1800 RPM, AC
Induction Motors

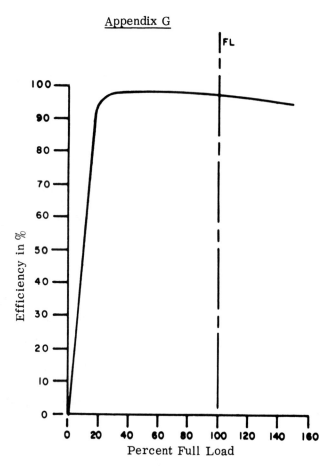

Appendix G

Efficiency Curve of 1000 HP DC Series or
Shunt Type Motor Used on Electric Drill Rig

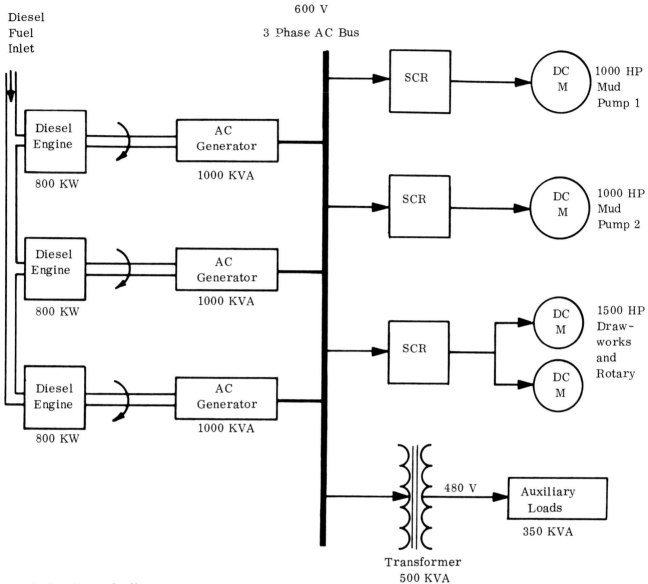

Diesel
Fuel
Inlet

600 V

3 Phase AC Bus

Typical Full Load Efficiences

Diesel Engine   –   32% – 34%

AC Generator   –   95%

SCR   –   99%

DC Motor   –   95%

Transformer   –   95%

A Typical SCR Rig Power System Flow Diagram

Figure 1

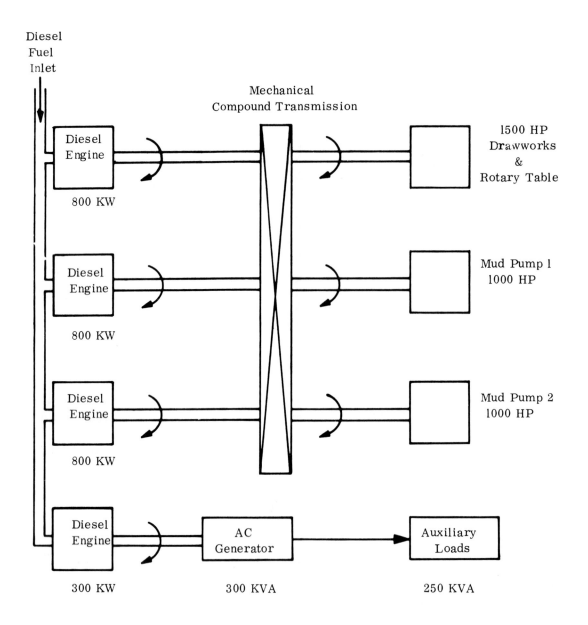

Diesel
Fuel
Inlet

Mechanical
Compound Transmission

Diesel
Engine

800 KW

1500 HP
Drawworks
&
Rotary Table

Diesel
Engine

800 KW

Mud Pump 1
1000 HP

Diesel
Engine

800 KW

Mud Pump 2
1000 HP

Diesel
Engine

300 KW

AC
Generator

300 KVA

Auxiliary
Loads

250 KVA

Typical Full Load Efficiencies

Diesel Engine   - 32% - 34%

AC Generator   - 95%

Mechanical Transmission - 75%

A Typical Mechanical Rig Power System Flow Diagram

Figure 2

Diesel
Fuel
Input

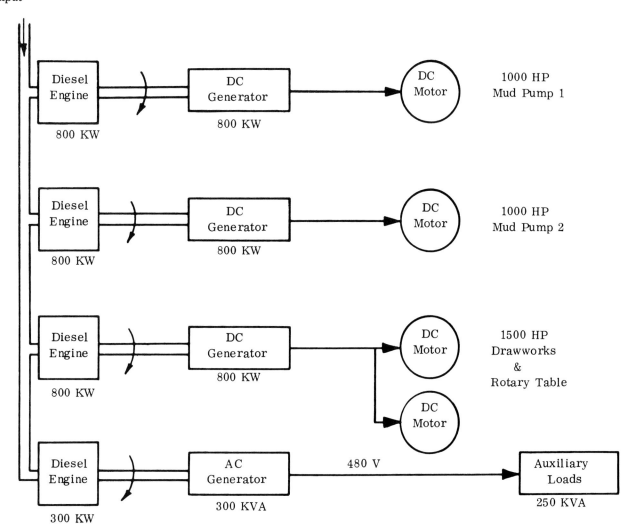

Typical Full Load Efficiencies

Diesel Engine	- 32% - 34%
DC Generator	- 95%
AC Generator	- 95%
DC Motor	- 95%

A Typical DC-DC Rig Power System Flow Diagram

Figure 3

Acknowledgements and References

1. McNair, Will L., _The Electric Drilling Rig Handbook_, Tulsa, Ok, Petroleum Publishing Co., 1980.

2. Norvell, A. B., _Diesel Engineering Handbook_, New York, Diesel Publications, Inc., 1955.

3. Parsons, Richard C., Santa Fe Drilling Co., "Drilling Cost Accounting Techniques", presented at 1979 Drilling Technology Conference of IADC, March 6-8.

4. Stone, Wm., Baylor Co., "A Systems Approach to Electric Land Rigs", presented at 1979 Drilling Technology Conference of IADC, March 6-8.

5. Waukesha Engine Div., Dresser Industries, Inc., Waukesha, Wisc., 1979. Engine Technical Data and Reports.

6. Webb, Glenn, General Electric Co., "A Comparison of Mechanical and Electric Drives for Land Drilling Rigs", presented at 1977 Drilling Technology Conference of IADC, March 16-18.

LESSON 2 QUESTIONS

Put the correct answer in the blank before each question. If there is more than one correct answer, put in all the correct letters. If a blank is drawn in the question, write out the answer as well as supply the letter in the multiple choice slot. The very act of writing down the answer will help you remember it.

Look again at Lesson 2 from the **Introduction** up to **Rig Design Considerations.** Answer the following questions:

_____ 1. Only the total weight of the rig is important when considering rig portability.
  A. True
  B. False

_____ 2. One of the requirements of rig design is to provide for good *hydraulics* by –
  A. allowing for rapid erection and take-down.
  B. providing the maximum amount of available power for the circulating fluid to reach the bit while drilling.
  C. letting drill pipe be run into the hole or pulled out with minimum time wasted.

_____ 3. Factors that determine a rig's portability include –
  A. whether certain units of equipment must be handled as separate packages.
  B. how many men will travel with the rig.
  C. the depth of hole the rig will be able to drill.

_____ 4. In the hoisting system, the _____ is powered by two or three prime movers to raise or lower the drill stem so that the bit can drill.
  A. kelly
  B. bit
  C. drawworks
  D. catheads

_____ 5. The weight of the drill stem on the hook and *elevators* (clamps that grip the pipe) is supported by the _____.
  A. casing
  B. catheads
  C. guy wires
  D. mast or derrick

_____ 6. Drilling is possible because the rotary table turns the _____ _____, which rotates the kelly.
  A. kelly bushing
  B. drill pipe
  C. swivel
  D. hook

_____ 7. The space in the borehole around the drill pipe is called the

_____.

    A. wall
    B. casing
    C. tubing
    D. annulus

_____ 8. The _____ power of the drilling fluid passing out of the bit washes rock cuttings to the surface.
    A. drilling
    B. jet
    C. hydraulic
    D. hoisting

_____ 9. A mast is assembled piece by piece at the drilling site.
    A. True
    B. False

_____ 10. The substructure supports the _____.
    A. mast
    B. blowout preventers
    C. rotary
    D. rig floor

_____ 11. Heavy-duty substructures will support a maximum of about 500,000 pounds.
    A. True
    B. False

_____ 12. One reason steam is no longer a source of rig power is that –
    A. electricity is more powerful.
    B. natural gas is more expensive than it used to be.
    C. nuclear power is commonly used.

_____ 13. Most drilling rigs today are _____.
    A. diesel-electric
    B. turbine-electric
    C. steam and mechanical

_____ 14. On a diesel-electric rig, the heaviest load of power is transmitted to the

_____.

    A. rotary table
    B. drawworks
    C. mud pumps

_____ 15. If power on a rig is transmitted through a compound, the rig is called _____

_____.
   A. an electric-drive rig.
   B. a mechanical-drive rig
   C. an internal-combustion engine

_____ 16. Which ones of the following parts might be found in the compound?
   A. Clutch
   B. Chains and sprockets
   C. Bit
   D. Pump

_____ 17. Most drawworks are equipped with two brake systems – one that is either electric

or hydraulic, and one that is _____.
   A. air-powered
   B. mechanical
   C. chained

_____ 18. The makeup cathead is located opposite the driller's console and is used to loosen
the pipe joints as the joints are pulled from the hole.
   A. True
   B. False

_____ 19. Another name for the pulleys in the crown block is _____.
   A. bearings
   B. cables
   C. sheaves

_____ 20. Wire rope is usually –
   A. silicon wire.
   B. multistrand steel cable.
   C. 1¼ to 1½ inches in diameter.

_____ 21. Casing is often heavier than the drill stem for the hoist to lift.
   A. True
   B. False

_____ 22. Ways to extend the life of wire rope include –
   A. lubrication.
   B. small-diameter sheaves.
   C. a program to move the wire rope.

_____ 23. The kelly is turned by the kelly bushing, which fits into the rotary table.
   A. True
   B. False

_____ 24. Functions of the rotary table include –
A. raising and lowering the kelly.
B. rotating the kelly.
C. supporting the slips.

_____ 25. A kelly may be all of the following except –
A. square.
B. round.
C. triangular.
D. six-sided.

_____ 26. Functions of the swivel include –
A. providing a passageway for the drilling mud to the drill pipe.
B. providing the power to make the kelly turn.
C. supporting the weight of the drill stem.

_____ 27. The hook hangs below the swivel and attaches to the kelly.
A. True
B. False

_____ 28. Circulation fluid is not always a liquid.
A. True
B. False

_____ 29. Which of the following correctly traces the path of the drilling mud?
A. Mud pit, standpipe hose, annulus, swivel, bit, drill stem, mud pit
B. Swivel, drill stem, mud pit, standpipe hose, annulus, pit, mud bit
C. Mud pit, standpipe hose, swivel, drill stem, bit, annulus, mud pit

_____ 30. The circulating fluid cleans the bottom of the hole by forcing the cuttings back into the formation.
A. True
B. False

_____ 31. If formation fluids begin to enter the borehole, the event is called a

_____.
A. blowout
B. blowout preventer
C. kick

_____ 32. Ram preventers go on the top and annular preventers on the bottom of a BOP stack.
A. True
B. False

58

_____ 33. In order to operate the BOP, hydraulic fluid is put under nitrogen gas pressure in

_____.

A. a pump
B. an accumulator
C. a tank

_____ 34. Which of the following would *not* be powered by the generator for the auxiliaries?
A. Shale shaker motors
B. Rig lighting
C. Mud pit stirrers
D. Mud pumps

_____ 35. An air compressor will usually send its air to _____

_____.

A. the rotary
B. pneumatic controls and clutches
C. the mud pumps

Look again at **Rig Design Considerations**. Answer the following questions:

_____ 36. Which of the following greatly influences the efficiencies of the other four systems?
A. Mud pump drives
B. Power transmission devices
C. Rotary drives
D. Drawworks and mechanical transmission
E. Wire rope system

_____ 37. With a mechanical-drive rig, a separate diesel-AC generator set must be used for electric lighting and auxiliary functions.
A. True
B. False

_____ 38. With a DC/DC drilling rig, a diesel engine drives an AC generator to power DC motors for the various drilling functions.
A. True
B. False

_____ 39. With an SCR electric drilling rig, a *silicon controlled rectifier* converts power from AC generators into direct current for the DC motors at the various drilling functions.
A. True
B. False

_____ 40. The SCR system must have a separate diesel-driven generator to provide electricity for the auxiliaries on the rig.
A. True
B. False

_____ 41. With the SCR system, several diesel-AC generator sets provide power to a common _____, where the SCR converters control the power supplied to the DC motors on the rig.
A. compound
B. clutch
C. AC bus

_____ 42. One of the virtues of the _____ is that a minimum number of generators can be run to supply only the needed amount of energy to the DC rig motors at any one time.
A. mechanical rig
B. DC/DC drive rig
C. SCR electric rig

_____ 43. The DC/DC system shows a _____ efficiency improvement over the SCR system, but the DC/DC system must also maintain an additional AC generator for the rig auxiliaries.
A. 1 percent
B. 3 percent
C. 5 percent

_____ 44. The most significant cost factor in operating a rig is the _____ _____.
A. type of generators used
B. cost of diesel fuel
C. maintenance costs

_____ 45. Both SCR electric rigs and DC/DC drive rigs have faster _____ _____ than mechanical-drive rigs.
A. rotary rates
B. personnel turnover
C. rig-up and tear-down time

_____ 46. A DC/DC drive system requires dedicated engine-generator sets for each DC motor for the special drilling functions on the rig.
A. True
B. False

_____ 47. In the case of a mechanical rig, when the drawworks is on the derrick floor the engines must be close to the same height to avoid long chain drives.
A. True
B. False

Look at figures 1–3 on pages 51–53 of the reprinted article for questions 48, 49, and 50. In the figures, *KVA* stands for kilovolt-amperes. The *800 KW* labels under the diesel engines mean that the diesel engines have the potential of producing 800 kW in the generator. (Actually, all the diesel engine does is rotate a shaft to power the generator.)

_____ 48. If only one mud pump were being used during drilling, how much horsepower would the SCR system be responsible for controlling?

A. 1,000

B. 350

C. 2,500

_____ 49. Why does figure 1 need only three diesel-generators and figure 3 need four?

A. Since the AC bus of an SCR system produces AC current, the auxiliary system can draw from the AC bus that is supplied by the three generators.

B. In a DC/DC rig design, the diesel-generators are dedicated to certain rig-function motors.

C. The auxiliary system needs more power than the three diesel-generators can produce.

_____ 50. Why does figure 2 need only one generator?

A. Since mechanical rigs are usually smaller, the one generator can produce enough power for all the drilling functions.

B. The drawworks, rotary table, and mud pumps are receiving their power in mechanical form from the compound transmission, and therefore do not use electricity.

C. All four diesel engines power the one generator so that it can produce more power.

Lesson 3
MUD CIRCULATION EQUIPMENT

Introduction

Mud Pits

Mud Mixers and Agitators

Mud Pit Instruments

Mud Storage and Handling

Mud Cleaning Equipment

Duplex Mud Pumps

Triplex Mud Pumps

Lesson 3
MUD CIRCULATION EQUIPMENT

INTRODUCTION

The main components of the fluid circulation system are the pump, hose and swivel, drill string, bit, mud return line, and the pits (fig. 3.1). Mud conditioning equipment includes shale shakers, mud agitators, desanders, desilters, mud centrifuges, mud-gas separators, and degassers. Accessory equipment to the mud circulation system includes the standpipe, chemical tank, mixing hopper, mud storage facilities, and mud pit instrumentation.

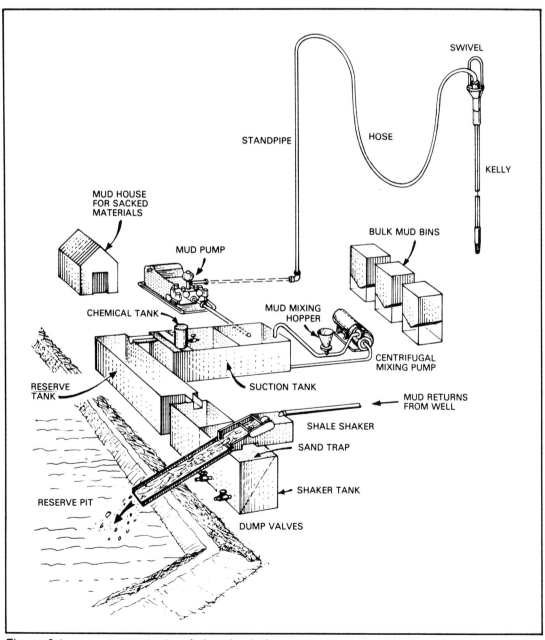

Figure 3.1. Main components of the circulating system

MUD PITS

The main function of the various mud pits and tanks on a drilling rig is to accumulate the mud circulated from the hole and to provide a constant supply of mud to the mud pump. The pits also serve as a reservoir for cooling and slowing the mud so that cuttings from the bottom of the hole can settle out before the mud is recirculated. Nearby equipment provides for conditioning the mud and mixing chemicals with it.

The *reserve pits* are used mainly as a depository for waste fluid, cuttings, and trash that accumulate as a well is drilled. To transfer fluid from the reserve pit to the active system, a small pump is used, or the retaining wall of the pit is opened to allow flow. Good mud is often partitioned off in a reserve pit called the *duck's nest*. Offshore rigs are usually not operated with reserve pits.

Earthen pits for circulation can sometimes be used for drilling medium-depth holes. Side-by-side pits may be used, but the end-to-end arrangement can be as effective. A short trench between the two pits ensures that the mud traverses the full length of each pit before reaching the mud pump. The slow movement of the mud through the pits permits cooling and settling. The first pit should be large enough to contain the sand and shale expected to be removed from the hole during drilling operations. The second pit should have enough mud to fill the hole when all the drill pipe, drill collars, and tool joints are out of the hole and also to leave a reasonable reserve in the pit. The settling pit can be smaller because the shale shaker at the outlet of the mud return line takes out large cuttings from the stream before it reaches the pit.

Figure 3.2. Steel mud tanks

Steel Tanks

A better arrangement for circulation through the pits uses steel tanks (fig. 3.2). This arrangement eliminates the expense of digging and boarding the side of earthen pits and the possible refilling of the pits once drilling is completed. Other advantages of steel tanks are that they have a known volume of mud at all times, allow easy access to cleaning the tanks, and provide the positive pressure needed for the suction of the pump. Chemical treating is also easier in steel tanks because mud volumes can be more accurately calculated. Mud-flow piping can be easily installed on steel pits and can be easily cleaned. Ditching may be arranged to permit the mud stream to bypass any tank and go directly to another. Assembly time is also saved because flexible connectors for mud pump suctions can be conveniently installed.

Mud tanks should be big enough to contain the mud volume that is required in the hole when the pipe is on bottom at total depth. A reasonable reserve of mud should also be in the pits to combat lost circulation and well kicks. Jets and pump-out connections should be installed at intervals on the bottom of the tanks so settled solids and sand can be removed. Agitators and stirrers should be provided to keep weighting material in suspension.

Arrangement

In the general arrangement of mud tanks and pits and mud conditioning equipment (fig. 3.3), the fluid flows through the shaker tank, settling tank, reserve mud tank, suction tank, and finally to the mud pumps.

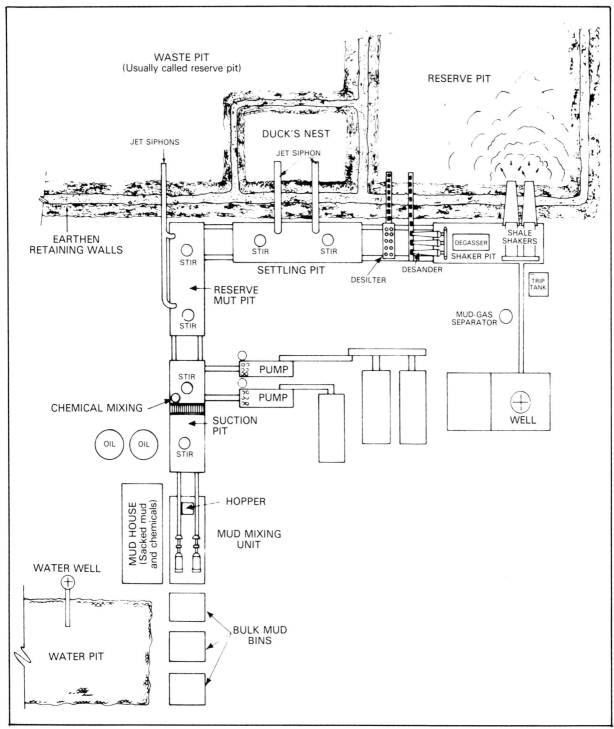

Figure 3.3. Mud pits and pumps for a modern drilling rig

The shaker tank usually has at least two compartments and one or more shale shakers. The compartment below the shale shakers acts as a sand trap. This compartment should have a sloping bottom and a jet siphon or dump valve for emptying sand and fine cuttings. The shaker tank usually has no mud-stirring devices. If a degasser is used, a common practice is to take suction out of the second compartment and return the degassed mud to the next tank. Similarly, if desanding and desilting equipment is used, it is operated downstream from the sand trap and the degasser. The degasser, desander, and desilter equipment is mounted on the top side of the unit with built-in piping for the pumps. Solids control equipment should always be placed in order of the size of particles removed.

The settling tank usually has two stirrers and mud guns for agitating the mud. Two or more jet siphons remove surplus mud that may accumulate in the tank. Top and bottom flow lines are provided to permit movement into and out of the tank. Dump valves are usually provided to allow emptying of the tank. The reserve mud tank is generally equipped in the same manner as the settling tank. In fact, the settling tank and the reserve tank serve the same functions – to permit settling time for the fluid and to provide storage room for conditioned mud that is in ready reserve.

A suction tank is usually provided with two compartments, one for the pump suctions and the other for a slug tank (pill tank), where quantities of heavy mud (slugs) can be mixed and conditioned. Slugs are pumped into the drill string before trips to empty the drilling fluid, preventing contamination of crew and rig floor. Mud-mixing pumps and hopper are housed on a skid. Mud and chemicals are stored in the mud house, close to the mixing unit.

Bulk mud bins may be provided for the barite used for weighting mud. Water for preparing drilling mud and for rig and equipment washdown may be stored in an earthen pit or in tanks. A water well is drilled to provide water on land locations. On offshore or inland water locations, water for drilling is obtained directly from the canal, bay, or sea.

MUD MIXERS AND AGITATORS

Jet Hoppers

Jet hoppers represent the accepted equipment for adding dry mud materials to liquid mud. Centrifugal pumps for low-pressure, high-volume mixing service are popular; they use 50- to 75-hp motors. A venturi tube downstream from the jet nozzle permits better mixing.

The mixing pump circulates fluid from the tank through the jet and back to the tank. The high velocity of the fluid through the nozzle lowers static pressure to below atmospheric. In effect, a vacuum is created, and thus material placed in the hopper is sucked into the fluid stream. Powdered mud materials such as clay, bentonite, barites, and chemicals, as well as solids, cellophane, nut hulls, and other pulverized material can be added to the mud through a jet hopper. The capacity of a hopper is relative to the size of the jet and the volume of fluid pumped through it. A small jet hopper can mix 200 to 400 pounds of weighting material per minute into the fluid stream, and larger units can handle twice that load.

Good results have been obtained with a 50-hp motor and a pump on one hopper (fig. 3.4). The venturi tube allows a better vacuum and thus a higher volume of solids into the fluid stream. Quicker dispersion also takes place with this arrangement. The butterfly valve should be closed when mixing is not going on to minimize aeration of the drilling fluid.

Mixing material may be introduced through a suction hose instead of a hopper. The hookup can be used to mix liquids by connecting the hose to the tank where oil or liquid chemicals are stored. This method is most effective in emulsifying oil or adding materials premixed in fresh water to a brine system.

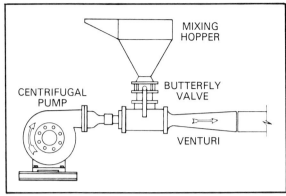

Figure 3.4. Centrifugal pump and jet hopper for mud mixing

Mud Guns

Mud flow through the pits can be as little as 2 feet per minute. Weighting materials will drop out of suspension unless vigorous stirring and agitation of the mud is provided. Mud guns serve this purpose on the drilling rig. They also break up high gel strength muds that form because of lack of stirring. A jet siphon used with a mud gun (fig. 3.5) results in good mixing of the drilling fluid.

Mud gun stands should be positioned around the pits so that the entire area is agitated. The guns are usually connected to a manifold and are supplied with mud from a pump for this purpose. The jet action of the guns gives the mud a swirling effect.

Centrifugal pumps are ideal for operating mud guns because of the large volume of flow that can be maintained. Variable flow can be easily regulated without causing excessive backpressure on the system.

Submerged mud gunning is preferred to permitting the mud stream to hit the fluid surface, which introduces air into the mud. Some submerged jets are swivel-mounted and thus permit the nozzles to rotate and roll the fluid in the mud pit.

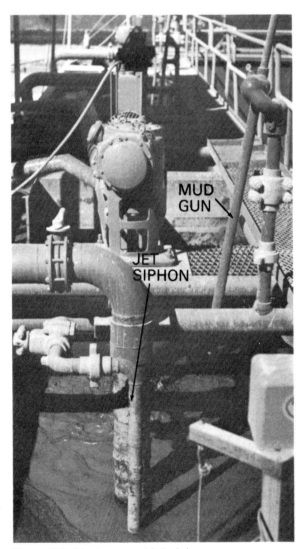

Figure 3.5. Mud gun and jet siphon

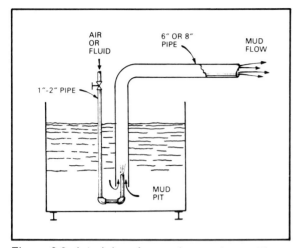
Figure 3.6. Jet siphon for moving mud or cuttings

Jet Siphons

A jet siphon (fig. 3.6) is used to transfer mud, sand, and cuttings from a mud tank to a reserve pit. Jet siphons originally used air or steam, but today use drilling fluid or water to suck fluid into the larger pipe to remove the fluid from the tank.

Figure 3.7, Propellor agitator powered by electric motor

Mud Agitators

Stirring mud in steel pits is done with electric propeller or paddle agitators (fig. 3.7). This equipment keeps the mud in movement and prevents gels from forming.

MUD PIT INSTRUMENTS

Mud pit (tank) instruments are not part of the mud conditioning system on a drilling rig. This equipment is involved with well pressure control so that kicks can be controlled and blowouts prevented. By knowing about pit instrumentation, the derrickman can be aware of warning signs of kicks so that the drilling fluid can be conditioned in time to gain control.

Pit Level Indicators

A pit level recording device like the pit volume totalizer (fig. 3.8) is used to indicate the total volume of mud in a pit. Air signals representing the actual number of barrels in each tank are sent to a special relay device, where the readings are averaged. The averaged signal is then transmitted to an indicator and recorder, where the totalized mud volume in the tanks is indicated on a chart. A sound alarm or warning lights can be triggered from the recorder. The driller can thus notice variations of as little as ½ barrel.

Pit volume totalizers (PVTs) are especially valuable in warning of potential well kicks or blowouts. They are also useful in determining the proper fill-up when pulling pipe out of the hole or when checking slow gains or losses in the mud volume.

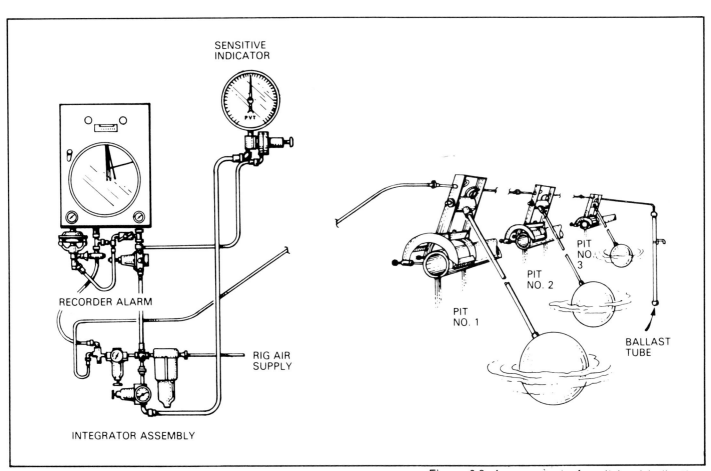

Figure 3.8. Arrangement of a pit level indicating instrument

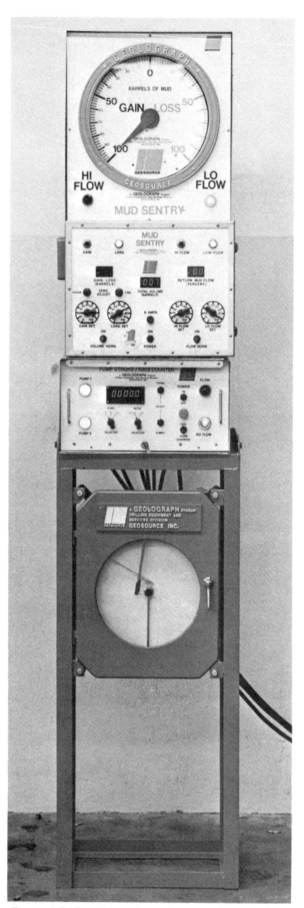

Figure 3.9. Electric mud sentry

The standard recorder device employs a 24-hour chart to show barrels, cubic metres, or any other unit of volume measurement. This chart helps the driller know about downhole conditions by noting changes as they occur.

Another type of pit level instrument operates on electrical signals instead of pneumatic controls (fig. 3.9). Pit level sensing elements are supported by floats and move vertically in the mud tanks to measure the level of hydrostatic pressure. Signals produced from each tank or compartment are processed by a totalizing relay and directed to the instrument with no more than ±2 percent error. The instrument may be combined with a flow sensor; both mud volume and flow are recorded on a chart.

Flow Sensor

A flow sensor is used to indicate mud flow from the return line of the well. The key element is a paddle that is mounted on the return flow line with a torque-indicating system. This system transmits a signal to an instrument on the rig floor, indicating well flow. The indication is independent of level changes in the mud tank. A flow sensor enables the driller to check pump efficiency, check hole fill-up during trips, spot fluids downhole, predetermine the number of strokes required before fluid or gas from bottom appears at the surface, and be certain the hole is full of mud at all times.

A pump stroke indicator is an essential part of the flow sensor. To check hole fill-up during trips, the instrument totals the number of mud pump strokes required to fill the hole after a given number of stands has been pulled. During the trip, the driller may determine if the hole is taking the proper amount of mud by comparing pump strokes required for the same number of stands with an original reference count. The paddle flow sensor can read mud flow within a percentage point of actual flow; gains or losses in mud flow can be accurately registered.

Pump Stroke Counter

Pump stroke counters enable the driller to quickly detect pump trouble and changes occurring in drilling conditions. A reciprocating mud pump is equivalent to a positive-displacement meter; thus a stroke counter is an accurate means of measuring the amount of fluid needed to fill the hole. Swabbing action occurring while pipe is being pulled may be detected by noting the mud volume required to fill the hole at regular intervals, usually every five or ten stands of drill pipe. Counters can be attached to the mud pumps to relay the number of strokes of each pump to the driller's location. Counters are frequently furnished as an added feature with other indicating devices.

Mud Weight Indicator

Mud weight indicators were developed primarily as safety devices for the prevention of blowouts. When mud density is critical, these indicators are useful for obtaining an accurate and continuous record of all mud pumped into the well. They may be air-actuated (fig. 3.10) or electrically actuated (fig. 3.11). An electric mud weight indicator offers a digital readout of *mud-in* and *mud-out* densities. The mud-in sensor is installed in the suction tank to measure the density before going downhole; the mud-out sensor is placed in the shale shaker box to detect any density changes in the mud returns. This instrument can also be arranged to read mud temperatures in and out of the well.

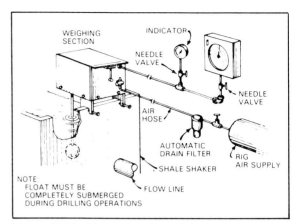

Figure 3.10. Air-actuated mud density recording device

Figure 3.11. Electrical mud density and temperature recording device (*Courtesy of Martin-Decker*)

MUD STORAGE AND HANDLING

Handling

Most mud materials are furnished in dry form, generally in sacks. Bentonite, clays, starches, CMC, lignins, ligno-sulfonates, polymers, and others are usually packaged in multilayer paper bags of 50 to 100 pounds. Barite weighting material is frequently handled in 100-pound bags, but much of this material is handled in bulk form. Barite is the only mud material that has proven to be advantageously handled in bulk form, both on land and offshore rigs.

Sacked Material

Truck transports usually bring sacked material to the rig, where it is stored in a mud house for protection from the environment. If small quantities and short storage time are involved, the sacks are stacked on wooden supports with canvas coverings for protection against rain. Mud houses (fig. 3.12) should be placed on an elevated foundation so that water accumulation on the ground does not damage the lower sacks. Sacked mud material is easily damaged if it is not kept dry. The mud house should be easily reached by delivery trucks, and mud sacks should be stacked for ease of handling. Broken sacks should be used at the earliest opportunity to minimize spillage losses. The mud house is usually located next to the mixing hopper and at the same level.

Bulk Mud

The advantages of handling mud materials in bulk form are—
(1) reduced labor requirements;
(2) increased speed of adding weight material to the mud stream;
(3) reduced losses from broken sacks; and
(4) cleanliness of handling, storing, and mixing operations.

Figure 3.12. Mud and chemical storage

From transport trucks, barite is loaded in the upper manhole of a rig storage tank. The barite is blown through a 4-inch hose to the storage facility. Pressurized air at the bottom of the tank works its way through the pulverized barite to the top of the container. The jetting effect caused by the air fluidizes the barite above the porous membrane or air slide. Air pressure above the material pushes the barite to the discharge pipe and then out the hose to the storage tank.

Modifications of truck transports are used on barges, supply boats, and other floating craft for moving pulverized material like cement on canals and to offshore locations. For offshore moves, a special pressure tank is provided to push the aerated material to the platforms, which may be 100 feet above the surface of the water.

Bulk barite is stored in rig tanks, or bins, on land locations. Bin capacities range from 500 to 1,000 sacks. The tapered bottom of the tank is equipped with an air slide and gate discharge that allows the fluidized barite to flow directly by gravity into a mixing hopper below the tank. Bulk tanks can mix weighting material at very high rates if the mud pumps can handle the volume needed to circulate the well. The actual rate of bulk barite mixing depends on the mixing pump and the fluid flow through the hopper. Weighting material added to fluid mud from bins can be measured by –

(1) mud weight gain in pounds per gallon and volume treated; and
(2) surface level of the dry mud in the bins.

When a drilling job is completed, the bulk mud remaining in the bins can be transferred back to the truck transport. This reverse process uses a blower and an air bazooka, a special aeration unit, to transfer the bulk material.

MUD CLEANING EQUIPMENT

Shale Shakers

The mud coming up the annulus from the bottom of the hole carries cuttings, sand, and other solid particles. This material must be removed from the drilling fluid before the fluid is recirculated into the well. This job requires high-capacity, vibrating screen shakers, or shale shakers (fig. 3.13). The mud passes over the shale shaker screens, where large solid particles are blocked from entering the mud stream. The mud that passes through the screen goes into the mud tank for further conditioning before it is pumped back into the well.

The material that remains on top of the shale shaker screen is the oversized material. Commonly used screens have relatively large openings. Since many fine particles may adhere to the larger coarse particles and could be discarded, spraying the screen with water should be avoided if the larger solids are to be discarded. A heavy spray of water on the screen may be too much to keep the mud in the desired condition.

Figure 3.13. Shale shakers

The sand trap compartment below the shale shaker is important. If the screen develops a hole, the sand trap stops larger particles from going downstream and possibly plugging mud conditioning equipment. If the shaker bypass is left open by mistake, the sand trap can catch and hold large particles for a short time. The trap may act as a desander below the vibrating screen when the mud is heavily loaded with sand.

The screen sizes customarily used in shale shalers in the top portion of the well – where hole size is large and the drilling rate is fast – permit sand and solids to settle quickly in most muds. To catch these solids, a mud tank with a *sand trap* is used. Cleanout openings should be provided to permit periodic dumping so that solids that have passed through the screen can be removed. The bottom of the sand trap compartment should be sloped so that separated particles can settle toward the openings. The sand trap should be cleaned so that coarse particles are discarded and the liquid mud remains.

Cement, cement-cut mud, used chemicals, and other contaminants should not pass over the shaker screen and enter the active mud system. Most shale shakers are provided with a swing-line or gate-and-trough arrangement that dumps contaminants into the waste pit.

Solids in the mud may be divided into two groups – low gravity (drill solids and bentonite) and high gravity (barite). The low-gravity solids are further subdivided into nonreactive and reactive groups. The *nonreactive solids* consist of sand, chert, limestone, dolomite, some shales, and mixtures of many minerals. These solids generally are undesirable and can be abrasive when larger than 15 microns. (A *micron* is one-millionth of a metre.) Table 3.1 gives screen sizes that will screen particles of various sizes.

Mud *reactive solids* are the clays. The term *reactive* describes the action of these solids in water. A clay may be defined as –

(1) a solid with an equivalent diameter of less than 2 microns;
(2) an electrically charged particle capable of adsorbing water; and
(3) a material that gives the appearance of swelling when water is adsorbed.

Two types of clays are generally used in drilling mud: (1) sodium montmorillonite, commonly called *bentonite* or *gel,* and (2) attapulgite, frequently called *salt gel.* Either of these materials can increase mud *viscosity* (resistance to flow) and cause treating problems if not controlled in the system. If clay solids are not removed from a weighted system, the usual result is an increase in viscosity. Adding water lowers the viscosity but also reduces the density. In turn, weighting material is added to raise the density to the desired figure. In a weighted mud, removal of undesired solids by mechanical

TABLE 3.1
SIZES OF SOLIDS AND SHAKER SCREEN

Solids (microns)	Solids (inches)	Shaker Screen
1,540	0.0606	12 × 12
1,230	0.0483	14 × 14
1,020	0.0403	16 × 16
920	0.0362	18 × 18
765	0.0303	20 × 20
		Laboratory Test Screens
210	0.00827	U.S.S. No. 60
147	0.00579	U.S.S. No. 100
74	0.00291	U.S.S. No. 200

means (such as a vibrating screen) is usually less expensive than chemical thinning or the addition of barite to maintain weight at a given figure.

A 12 × 12 mesh shaker screens out ¹⁄₁₆-inch particles. A 20 × 20 mesh screen removes 0.03-inch particles. Table 3.2 shows how one fine-screen shaker screens out particles using various mesh screens. Note the reduced mud flow capacity across the shaker unit when the screen mesh size is reduced.

Fine screening presents special problems that must be recognized. Normal shaker screens are made of heavy wires and have an open area of 50 to 55 percent, whereas an 80-mesh screen is made with much finer wire that is not as rugged and has an open area of only about 30 percent. To handle the same volume of fluid, the finer mesh screen has to be about 50 percent larger in total surface area. For this reason, it is impractical to use fine-mesh screens with the high circulating volumes normally employed with top-hole drilling. However, fine-mesh screening is both practical and desirable for intermediate and deep-hole drilling because the mud becomes easier to maintain in good condition with greater removal of undesirable solids. The cost for mud chemicals is less, and the overall cost of operation is lower when fine solids are not left in the mud.

The type of mud used affects screening capacity. Oil-base muds generally have higher normal viscosities than do water-base muds of the same weight. The ability of fine screens to handle oil muds is reduced accordingly. Mud components, such as some of the synthetic polymers or finely divided asbestos, tend to reduce screen capacity.

Maintaining vibrating-screen shale shakers in good condition is relatively simple. Bearings should be greased daily. Tension bolts that hold the screen in place should be checked regularly for tightness. The screens should be kept pushed all the way to the back of the screen boxes to keep solids from leaking. If only one screen is used on a double-deck shaker, it should be placed on bottom. The condition and tension of the drive belt should be checked weekly. The bypass gate and flow directors should be checked occasionally to make sure that they are free to move.

Desanders and Desilters

A cross-sectional diagram of a cone-shaped centrifugal separator is shown in figure 3.14. This equipment has no moving parts, but imparts a whirling motion to the drilling fluid with enough force to separate various-sized particles. It is a simple and inexpensive machine to operate and has a relatively high capacity. A centrifugal pump is used to feed mud through a side opening into the large end of the cone-shaped housing. The fluid whirls much like a water spout or tornado. All cone-shaped separators operate in a similar manner, whether used as desanders, desilters, or for the recovery of weighting materials.

TABLE 3.2
SCREENING FOR VARIOUS PARTICLE SIZES

Screen Mesh	Micron Size	Particle Diameter (inches)	Mud Flow, Single Screen Unit (gpm)
40	1,190	0.047	800
60	250	0.010	600
80	177	0.007	400
100	149	0.006	300

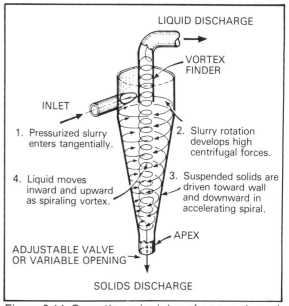

Figure 3.14. Operating principles of a cone-shaped centrifuge

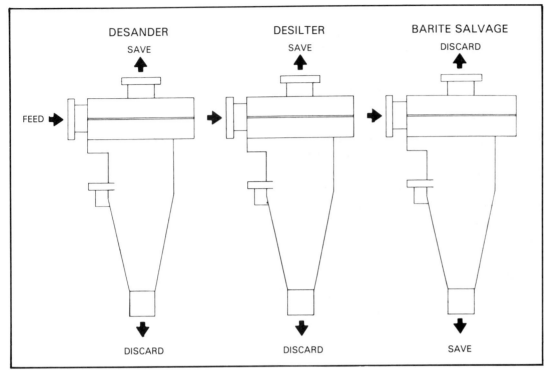

Figure 3.15. Applications of a cone-shaped centri-fugal separator

Figure 3.16. Desander

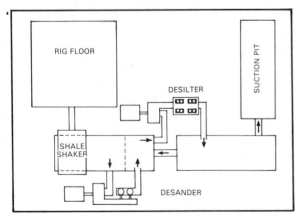

Figure 3.17. Recommended desander and desilter installations

In figure 3.15, three applications of cone-shaped separators are illustrated. When a cone unit is being used as a desander or desilter, the underflow from the apex contains the coarse solids, which are discarded, while the overflow is returned to the mud stream. When a cone unit is used on weighted muds, the underflow contains barite, which is saved and returned to the mud stream.

Desanding cones have an internal diameter of 6 to 12 inches. They have the advantage of handling large volumes, but they do not discharge as many solids as smaller cones.

Desilters (fig. 3.16) have 3-inch, 4-inch, or 5-inch cones. They are capable of rejecting extremely fine particles. All sizes of sand, a high percentage of materials larger than 10 to 20 microns, and even material down to 2 to 3 microns can be removed by a desilter. Total and constant desilting of the mud system can drastically reduce mud pump wear, hole problems, the number of bits, the time required to drill the hole, and water and chemicals required to treat the mud.

Desanders and desilters should be installed and operated early in a drilling program. The desilter should be located so that mud can be picked up downstream of the shale shaker and discharged into the next tank (fig. 3.17). The desander unit is placed above the mud tanks to facilitate disposal of the underflow.

Mud Centrifuges

In solids control of weighted muds, the object is to discard particles larger than barite by screening, and particles smaller than barite by centrifuging. Barite, which is silt size, is left in the mud. Unavoidably, drilled solids that are silt size are also left.

A *decanting mud centrifuge* is primarily used with weighted muds to recover weighting materials and reject materials that increase viscosity. The principle of the centrifuge is illustrated in figure 3.18. This machine is efficient and is capable of cutting fine material between 2 and 5 microns, depending on the specific gravity of the fluid solids. Particles larger than 5 microns are separated into one stream, and those smaller than 2 microns are separated into another stream. A decanting centrifuge is sometimes used to remove drilled solids from unweighted oil-base muds, since it can discard the coarse solids as a semidry sludge with minimum loss of the valuable liquid phase.

When a weighted mud is being processed to remove and save the barite, the mud is diluted with water as it enters the machine in order to reduce viscosity. About 3 gallons of water per minute is sufficient. As the diluted slurry starts to rotate, the larger and heavier particles are thrown to the outside of the bowl and are conveyed to the coarse-solids discharge. These solids contain the bulk of the barite from the feed mud, and they are returned to the active mud system. The liquid fraction retains the viscosity-producing fine solids, which are discarded.

Figure 3.18. Principle of the decanting centrifuge

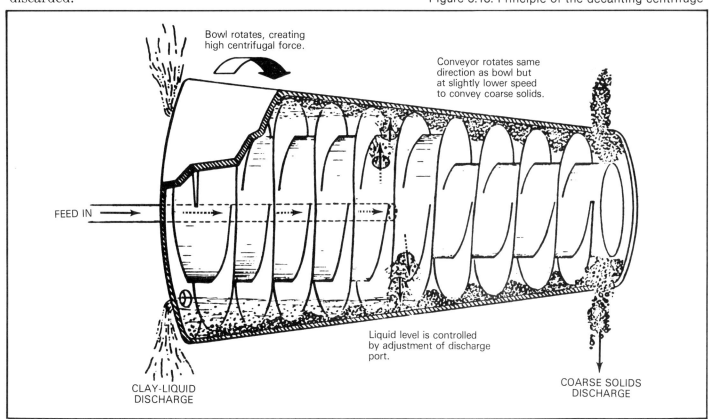

Bowl rotates, creating high centrifugal force.

Conveyor rotates same direction as bowl but at slightly lower speed to convey coarse solids.

FEED IN

Liquid level is controlled by adjustment of discharge port.

CLAY-LIQUID DISCHARGE

COARSE SOLIDS DISCHARGE

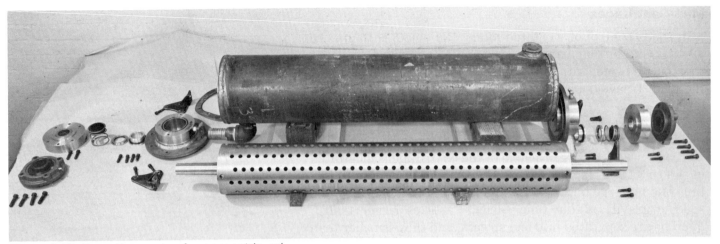

Figure 3.19. Operating parts of a concentric cylinder centrifuge

Because it makes cuts at 2 to 5 microns, the decanting centrifuge allows many of the coarser drilled solids and sand particles to mix with the barites returning to the system. The equipment is most efficient for barite recovery, and can recover 97 to 99 percent of barite particles larger than 2 microns.

A *concentric centrifuge* for recovery of barites separates the mud fluid into two streams, one rich in barite, the other containing the rest of the barites and the clay, silt, and chemical-bearing portions of the mud. This type of centrifuge (fig. 3.19) is composed of a stationary outer case of 8-inch pipe, 4-5 feet long, inside of which is a tube with a slightly smaller diameter, containing perforations. This tube revolves concentrically. Two positive-displacement pumps supply mud and water to the annulus (space) between the case and the rotor. In the annulus, the spinning rotor causes the mud to rotate. Centrifugal force moves the solids toward the case wall. The fluid next to the rotor surface loses barite. At the same time, the underflow pump is removing the barite-rich fluid at the case wall from the exit end of the separator at a specified rate. This rate is always less than the rate at which fluid is being pumped into the centrifuge. This imbalance in the flow causes the barite-depleted fluid at the rotor surface to pass through the holes in the rotor and out of the separator. Depending upon the purpose for which a centrifuge is used, the barite recovery rate may range from 70 to 95 percent. Barite recovery must justify the expense of the centrifuge.

Degassers

One problem with drilling fluids is *gas-cut mud*. Fluids can become gas-cut during the drilling of gas-bearing formations. Insufficient hydrostatic pressure of the mud in a well, in which case formation gas may enter the hole, can also cause gas-cut mud. In such cases, mud circulated from bottom may be several pounds per gallon lighter than the fluid pumped into the hole. Gas and mud flowing from a well when it is attempting to blow out can be dangerous. The gas must be vented a safe distance from the well, and the liquid mud should be returned to the pits to prevent waste. Recirculation of gas-cut mud is hazardous and can result in reduced pumping efficiency and lower hydrostatic pressure. Regular methods of using a vibrating screen and mud guns may not be effective, and mud-gas separators or vacuum degassers may have to be used.

A mud-gas separator is desirable to safely handle high-pressure gas and well flows. A vacuum degasser is more appropriate for separating entrained gas, which appears as foamy mud. A vacuum degasser speeds the settling process and the separation of mud and gas.

Mud-gas separators come in a variety of forms. Most consist of a vertical vessel arranged to vent gas from the upper end and discharge relatively gas-free mud from the bottom. Mud and gas from the well enter the unit through a line at the top. Free gas from the mud is taken from the separator and piped to a point for safe venting. Liquid mud is taken from the unit from a connection near the lower end, then through a riser and horizontal line to the shale shaker and pits. Working pressure of the unit may be as much as 100 psi.

A mud-gas separator arrangement utilizing a production vessel (fig. 3.20) may be mounted on an elevated structure, permitting gravity flow of fluid to the shale shaker. A pipeline allows venting of the gas at a safe distance from the well.

Figure 3.20. Production separator in use as a mud-gas separator

The primary use of a separator is to vent trip gas and to free much of the gas from heavily gas-cut mud. As gas slugs rise to the surface, their rapid expansion gives tremendous velocity to the mud ahead. If not controlled, this action can cause the mud to be blown out of the pits, damage rig equipment, and cause injuries. Gas following the mud creates a dangerous fire hazard when it escapes close to the rig.

During operation of a mud-gas separator, the hole must be shut in and mud circulated through the choke manifold; well flow is diverted from the flow line or choke manifold to the mud-gas separator. The mud and gas flowing into the separator release gas, which is carried by the vent line at the top to a remote flare. Moderate back-pressure can be carried on the separator, thus reducing the rate of expansion of gas in the mud stream and moderating the surge effect as slugs of gas and mud reach the surface. The mud separates from the gas and accumulates at the bottom of the separator. A float valve arrangement maintains a minimum fluid level in the vessel to prevent gas from blowing out the mud return line leading to the pit.

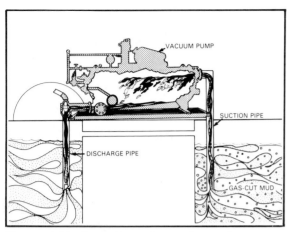

Figure 3.21. Operation of one type of vacuum degasser

One popular type of *vacuum degasser* (fig. 3.21) is mounted over a mud tank from which it takes suction. Mud enters the degasser through an 8-inch suction pipe because vacuum is applied by the vacuum pump mounted on the unit. The mud enters near the top of the horizontal barrel and flows along a section of large pipe that is closed at one end. The top of the pipe is cut horizontally so that mud can spill over the sides and down an incline extending the full length of the feed pipe.

As the mud flows down in a thin stream, the vacuum in the vapor space causes the gases to leave the mud and to be withdrawn from the tank. The degassed mud, which is back to its normal weight, flows to the bottom of the vessel for exit. The mud flows through the bottom of the vessel through the tube at the left of the machine into the second mud tank. A hydraulic jet in this downspout pumps mud at high velocity to lower the mud pressure below that in the degasser. This action causes the mud to flow from the degasser in spite of the vacuum. Safety devices prevent mud from entering the vacuum pump, and washdown lines for cleaning prevent plugging.

DUPLEX MUD PUMPS

The mud pump is the primary component of any fluid circulating system. It provides the driving force that sends the fluid through the route that it must travel. Almost all mud pumps today are powered by diesel engines, gas engines, or electric motors. Those designed for drilling have ratings up to 1,750 input horsepower (hp). They are capable of moving large volumes of fluid at pressures as high as 5,000 pounds per square inch (psi). They are often *duplex,* double-acting, reciprocating types; but *triplex,* single-acting pumps have become popular in recent years. *Centrifugal* pumps are used

Figure 3.22. Duplex, double-acting mud pump

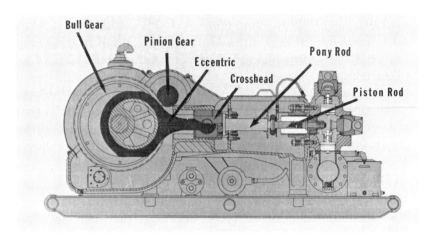

Figure 3.23. Power end, *left*, and fluid end, *right*, of a mud pump

for general service. *General service* includes supplying water for washdown, cooling brakes, mixing mud, supercharging the mud pumps, and circulating mud through the various conditioning devices.

A duplex pump (fig. 3.22) is so called because it has two cylinders, each containing a piston that moves back and forth inside a liner. One end of the pump, the fluid end, is concerned with intake and output of the fluid being pumped (fig. 3.23).

Each cylinder is attached to two intake valves and two discharge valves (fig. 3.24). One intake valve and one discharge valve are located on either end of the stroke area of the piston. The movement of the piston causes the valves to act in pairs; both an intake of mud into the pump and a discharge of mud from the pump occur during each back-and-forth movement of the piston. For this reason, the pump is described as double-acting.

As mud enters the suction line of a pump and flows through the pump, its velocity is continually changing due to the pulsating flow characteristic of a reciprocating pump. The

Figure 3.24. Intake and discharge valves of a duplex pump

83

double-acting duplex pump employs four equal-length strokes, each timed for 90 degrees of crankshaft rotation from the end of one stroke of one piston to the start of the next stroke of the other piston. Four pressure pulses per revolution of the crank are produced.

Because more fluid occupies the head side of the piston than the rod side, more fluid is moved and greater pressure pulses occur when the head sides of the pistons are being displaced. These pressure pulsations are made greater by–

(1) loss of effective suction head;
(2) fluid or hydraulic knocking;
(3) reduction of volumetric efficiency; and
(4) discharge line vibrations.

Effective suction head is reduced by a long suction line (because of greater fluid friction losses before the fluid reaches the piston) and by any suction arrangement that requires high lifting of the fluid. The pump and its suction should be hooked up to provide as much positive head as possible. The usual rig arrangement is to use mud tanks to provide a flooded suction.

Fluid *knocking* is closely related to insufficient suction head. Fluid knock results when the incoming mud cannot keep up with the piston as it speeds up during the stroke. At some point in the piston stroke, the fluid catches up with the piston, and an audible knock results. The degree of such fluid knocking depends on the conditions of the pump suction. Fluid knock should be avoided whenever possible.

Output

Volumetric efficiency (the percentage of fluid behind the piston on each stroke compared to the fluid that *could* occupy the space swept by the piston) is a measure of the effectiveness of the pump. If the piston area of a certain pump is 30 square inches and the stroke length is 16 inches, 480 cubic inches are swept on each separate stroke of the pistons (30 × 16). If only 408 cubic inches of fluid follows the piston and is later discharged, then the volumetric efficiency of the pump is said to be 408/480 or 85 percent.

Gas- or air-cut mud is the most frequent cause of reduced volumetric efficiency and substantially reduces the effective capacity of any pump. Leakage between piston and liner, liner and pump body, or valve and seat will result in loss of volumetric efficiency. Fluid cutting is usually severe before pressure reduction or loss of volumetric efficiency is noticed.

Most rig pumps run at essentially constant speed for a given engine throttle setting. When the load on the pump for a given liner size becomes too great, a smaller liner must be used. Output volume is less for a smaller liner than for larger sizes at the same speed. Table 3.3 shows the amount of fluid that duplex pumps deliver with various liner sizes and stroke lengths (at 60 spm). For volumes at other speeds, the amount of fluid is divided by 60 and multiplied by actual strokes per minute

TABLE 3.3
AVERAGE PUMPING RATES OF DUPLEX PUMPS AT 60 SPM
(GALLONS PER MINUTE)

STROKE (INCHES)	LINER DIAMETER (INCHES)				
	7	6½	6	5½	5
18	664	565	472	388	311
16	599	511	429	354	286
14	531	454	382	316	257

84

(spm). This table is based on 100 percent volumetric efficiency and average piston rod diameters. Few duplex pumps achieve much more than 90 percent volumetric efficiency in mud service because of the suction arrangements, pump speed, weighted fluids, and other conditions. High-speed operation results in lower volumetric efficiency than can be obtained at lower speeds. (Modern high-power pumps have larger rods than those shown in table 3.3, and thus pump somewhat smaller volumes of fluids.)

Different pumps may have different pressure ratings for the same liner diameter, depending on their mechanical design and input horsepower rating. Pressure ratings are determined by the load on the piston rod, which is the product of the pressure multiplied by the area of the piston exposed to fluid pressure. Pump nameplate ratings for various sizes of liners must be consulted, and specified pressures for given sizes of liners must never be exceeded. The liner size, which is the piston diameter, is critical to the pressure and volume output of any reciprocating pump. With large liners, low-pressure, high-volume output is obtained. Conversely, small liners permit high-pressure, low-volume fluid delivery. Normally, the size of the liner limits the allowable discharge pressure, regardless of speed. Reciprocating pumps require maximum horsepower input to operate at full rated speed, but it is often desirable to operate the pump at something less than top speed.

TRIPLEX MUD PUMPS

Triplex single-acting pumps have been used in the oil industry for many years, mainly for high-pressure acidizing, cementing, workover service, and pumping water. It was not until the late 1950s that single-acting plunger pumps, in both triplex and six-cylinder configurations, were put into mud service.

Duplex double-acting pumps were used almost exclusively for rig service until about 1960. As these pumps became larger—up to 1,500 hp—weight got to be a real problem. Plunger pumps are lighter in weight per horsepower and can be operated in high-pressure service with smooth discharge flow. Plunger pumps, of course, are inherently single acting in that only one pressure surge takes place with each forward stroke of the plunger. In the case of a triplex, fluid delivery strokes take place at each 120 degrees of crankshaft rotation; thus, three pressure surges are produced for each revolution of the crank. Because of excessive packing wear—due to the problem of cooling the plunger—the plunger design was abandoned in favor of half pistons and liners, as used in the double-acting duplex pump. Thus the single-acting triplex pump was accepted for heavy-duty, high-pressure drilling service. Single-acting piston pumps are commonly operated at 120 to 160 spm, compared to 60 to 70 spm for duplex pumps.

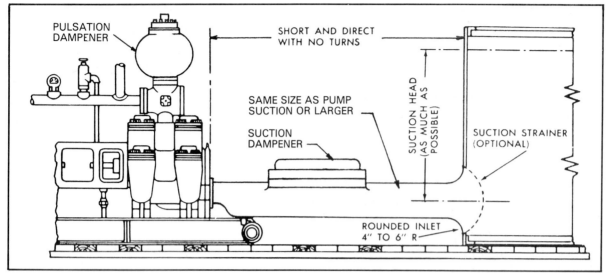

Figure 3.25. Pump suctions should be large, short, and straight.

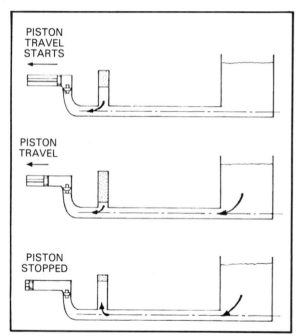

Figure 3.26. Operation of a suction pulsation dampener

Double-Action Pump Operation

The suction line from the mud tank to the pump should be large, short, and straight (fig. 3.25). Changes in size and direction of the line should be avoided when possible to keep losses to a minimum. Valves should open fully and be the same size as or larger than the suction line of the pump. A suction pulsation dampener, or desurger (fig. 3.26), reduces hydraulic hammer. The opening between the suction line and the dampener should be as large as practical. Any restriction here reduces the effectiveness of the dampener.

A pulsation dampener absorbs pressure variations, reduces peak pressures, permits high pump output, and minimizes damage to the discharge line, pump, and pump parts. When correctly utilized, a pulsation dampener effectively smoothes out discharge pressure variations. The dampeners should be as near the pump as possible. Nitrogen pressure must be in accordance with the manufacturer's recommendations.

Triplex single-acting piston pumps are 40 to 50 percent lighter in weight than duplex pumps having the same rated horsepower. Fluid end parts are smaller and more accessible than comparable parts in double-acting units. Various liner sizes for duplex and triplex pumps are shown in table 3.4. Note that liners for the triplex are much lighter, based on average weights in each instance. Piston rods seldom require replacement on a triplex pump. When needed, replacement parts for the triplex cost less than half the price of those for a duplex pump. Liners and pistons can be replaced in a triplex pump in less than half the time needed for a duplex. No rod packing and no stuffing boxes are used in a triplex unit. Triplex units can operate at higher pressures and produce smoother flow than is possible in duplex pumps. Some triplex pumps for drilling service are satisfactory for 5,000-psi maximum pressure.

TABLE 3.4
COMPARATIVE WEIGHTS OF PUMP LINERS

LINER SIZE (INCHES)	WEIGHT OF LINER (POUNDS)			
	Duplex (500 hp)	Triplex (500 hp)	Duplex (1,000 hp)	Triplex (1,000 hp)
4	338	98	–	119
5	313	81	466	98
6	259	55	394	67

Fluid delivery of triplex pumps–just as in duplex units–varies directly according to the speed of operation. Short-stroke triplex pumps are usually rated at higher speeds than the longer-stroke units. At a constant rate of strokes per minute, fluid delivery for a given stroke length is greater for large-diameter liners. The variation in fluid delivery with liner size and stroke length at 100 percent volumetric efficiency is shown in table 3.5.

Operation

The open bore of the power end of the liners in triplex pumps must be washed with water or soluble oil. Clean, cool water is the most economical. Experience has shown that, for the usual triplex pump in drilling service, 5 to 10 gallons of fluid per minute on each liner are needed to wash, cool, and give the best life for pistons and liners. Without sufficient flushing fluid on the piston and liner, the piston burns and scores the liner in less than 30 minutes of operation.

Since the triplex pump is a high-speed pump, it should have a charging pump for the suction. A charging pump has definite advantages, since it will ensure total filling of the pump cylinders at any speed. It is a necessity for satisfactory pump operation at higher operating speeds. The major cause of poor operation of a triplex pump is improper suction and discharge piping. Excessive pressure pulsations or surges in the suction system of a pump are the result of uneven or insufficient filling of the fluid cylinders. When insufficient filling of the fluid cylinder occurs, the pump, pump parts, and the discharge line of the pump are damaged due to shock loads. Poor suction condition (i.e., not filling the cylinder completely) produces poor volumetric efficiency and, in some cases, a fluid knock.

The primary purpose of a charging pump is to keep the mud pump fluid end full of mud at all times by maintaining a positive pressure in the suction line. Keeping the fluid end filled eliminates the impact loading of the pistons and the slamming action of the valves, and in turn eliminates the cause of shock loading on the power-end bearings and gears.

A very high pressure at the mud pump suction may cause sluggish valve action. On the other hand, a very low pressure at the mud pump suction will not eliminate shock and fluid hammer. Therefore, the discharge pressure of a supercharging pump must be carefully set and maintained.

TABLE 3.5
NOMINAL PUMPING RATES OF TRIPLEX
SINGLE-ACTING PUMPS AT 130 SPM
(GALLONS PER MINUTE)

STROKE (INCHES)	LINER SIZES (INCHES)				
	7	6½	6	5½	5
8	–	–	382	321	265
9	–	–	429	360	297
11	715	616	525	441	–

LESSON 3 QUESTIONS

Put the correct answer in the blank before each question. If there is more than one correct answer, put in all the correct letters. If a blank is drawn in the question, write out the answer as well as supply the letter in the multiple choice slot. The very act of writing down the answer will help you remember it.

Look again at the **Introduction, Mud Pits, Mud Mixers and Agitators, Mud Pit Instruments, Mud Storage and Handling,** and **Mud Cleaning Equipment**.

_____ 1. The fluid circulation system includes –
  A. mud pump.
  B. hose and swivel.
  C. drill string.
  D. mud return line.
  E. all of the above.

_____ 2. Mud conditioning equipment includes –
  A. shale shakers.
  B. mud return line.
  C. desilters.
  D. mud pumps.
  E. all of the above.

_____ 3. The reserve pit for an offshore rig is called _____
      _____
  A. the mud repository
  B. the duck's nest
  C. the fish bowl
  D. none of the above

_____ 4. The functions of the mud pits are –
  A. to settle the sand and shale removed from the hole.
  B. to cool the mud.
  C. to hold mud needed downhole when the drill stem is removed.
  D. all of the above.

_____ 5. Study figure 3.1. Mud flows through the fluid circulation system components as follows:
  A. kelly, swivel, standpipe, mud pump, suction tank, settling tank.
  B. suction tank, shaker tank, settling tank, mixing pump, standpipe, kelly.
  C. shaker tank, settling tank, suction tank, mud pump, standpipe, hose, kelly.
  D. kelly, standpipe, settling tank, reserve pit, suction tank, mud pump.

_____ 6. Functions of the settling tank are –
   A. to permit settling time for the fluid.
   B. to provide storage room for conditioned mud.
   C. to mix in barite to weight the mud.
   D. the same as that of the reserve tank.
   E. all of the above.

_____ 7. Slugs of heavy mud are pumped into the drill string –
   A. before tripping out, to prevent spilling mud on the crew and rig floor.
   B. after changing the bit, to clean out cuttings.
   C. during a connection, to lubricate the joints.

_____ 8. On offshore rigs, mud is mixed using stored fresh water from supply ships.
   A. True
   B. False

_____ 9. The jet hopper –
   A. adds dry mud materials to liquid mud.
   B. is the helicopter used to transport workers to offshore rigs.
   C. aerates drilling fluid.
   D. does none of the above.

_____ 10. Centrifugal pumps are well suited for operating mud guns because –
   A. they produce very high pressures.
   B. they have no moving parts.
   C. they pump large volumes of fluids.
   D. they do not break down.

_____ 11. Jet siphons –
   A. are powered by gas turbine engines.
   B. are useful for removing mud from a reserve tank.
   C. have no moving parts.
   D. remove jet fuel from petroleum.

_____ 12. Which of the following is not a function of the fluid circulating system?
   A. Agitating the drilling mud to break up gels
   B. Holding enough mud in reserve to replace the volume of the drill string while tripping out
   C. Mixing slugs of weighted mud to control a kick
   D. Storing oil produced from a reservoir formation
   E. All of the above

_____ 13. Pit volume indicators are valuable for –
   A. determining the weight of the mud coming out of the hole.
   B. warning of potential blowouts.
   C. determining how much mud is needed when pipe is pulled out of the hole.
   D. checking slow gains or losses in the mud volume.
   E. all of the above purposes.

_____ 14. The flow sensor operates by measuring changes in mud tank levels.
  A. True
  B. False

_____ 15. A flow sensor enables the driller to –
  A. check pump efficiency.
  B. measure formation volume.
  C. check hole fill-up during trips.
  D. be certain the hole is full of mud.
  E. do all of the above.

_____ 16. The number of strokes of a reciprocating mud pump accurately measures –
  A. the total volume of mud in the circulating system.
  B. the total weight of mud in the circulating system.
  C. the total volume of mud pumped over a measured period.
  D. the total weight of mud pumped over a measured period.
  E. all of the above.

_____ 17. Having to add less than the expected amount of mud while tripping out indicates loss of fluid to the formation.
  A. True
  B. False

_____ 18. The following statement is true of mud weight indicators:
  A. They measure densities of mud going into and coming out of the well.
  B. They may be pneumatically or electrically actuated.
  C. They were developed to help prevent blowouts.
  D. They may be used to measure mud temperatures electrically.
  E. All of the above statements are true.

_____ 19. Most mud materials are handled in bulk form.
  A. True
  B. False

_____ 20. Sacked mud materials are stored in _____.
  A. the doghouse
  B. the cathead
  C. the mud house
  D. a reserve tank
  E. none of the above

_____ 21. One of the following is *not* an advantage of handling dry mud materials in bulk form:
  A. Elimination of need for measurement
  B. Reduced labor requirement
  C. Reduced loss from broken sacks
  D. Increased speed and cleanliness of handling

_____ 22. Barite is moved into and out of the storage tank by _____

_____.

    A. gravity
    B. water
    C. pressurized air
    D. mud pumps
    E. electrostatic attraction

_____ 23. Weighting material added to fluid mud from bins can be measured by –
    A. fluid weight and volume gain.
    B. fluid volume gain alone.
    C. fluid loss to formation.
    D. level of material in bins.
    E. any of the above.

_____ 24. The shale shaker –
    A. adds shale to the drilling mud.
    B. vibrates.
    C. removes cuttings from drilling fluid coming out of the hole.
    D. mixes dry materials into the drilling fluid.
    E. does all of the above.

_____ 25. The sand trap –
    A. catches large particles which may get past the shale shaker.
    B. removes sand from sand-loaded cuttings.
    C. traps excess cement when cementing casing.
    D. separates high-gravity mud solids from low-gravity solids.
    E. does all of the above.

_____ 26. Bentonite is a reactive low-gravity solid.
    A. True
    B. False

_____ 27. A clay is –
    A. a solid with an equivalent diameter of less than 2 microns.
    B. a material that appears to swell when adsorbing water.
    C. an electrically charged particle capable of adsorbing water.
    D. all of the above.
    E. none of the above.

_____ 28. Why is adding water not the best way of reducing mud viscosity?
    A. It also reduces the density.
    B. It is always lost to the formation.
    C. It often requires the addition of expensive weighting components.
    D. It increases screen corrosion.

_____ 29. Coarse-mesh screens must be larger than fine-mesh screens to handle mud flow.
   A. True
   B. False

_____ 30. In a cone-shaped centrifugal separator, coarse sediments are separated from the fluid and removed through the apex.
   A. True
   B. False

_____ 31. Desilting –
   A. increases circulation loss.
   B. reduces bit wear.
   C. increases drilling efficiency.
   D. reduces mud pump wear.
   E. does all of the above.

_____ 32. A decanting mud centrifuge –
   A. removes particles larger than barite.
   B. agitates and mixes the mud to prevent formation of gels.
   C. removes barite-size drilled solids.
   D. removes particles smaller than barite.
   E. does all of the above.

_____ 33. The coarse solids discharged from the decanting mud centrifuge are returned to the active mud system.
   A. True
   B. False

_____ 34. In a concentric-centrifuge separator, solids move toward the –
   A. center.
   B. case wall.
   C. lower-pressure area.
   D. none of the above.

_____ 35. Gas-cut mud may be caused by –
   A. excessive aeration in the mud tanks.
   B. chemical reactions among mud components.
   C. insufficient hydrostatic pressure of mud in a well.
   D. drilling through gas-bearing formations.
   E. all of the above.

_____ 36. Recirculation of gas-cut mud can result in –
   A. dangerously low hydrostatic pressure.
   B. settling out of solid particles larger than 8 microns.
   C. excessive bit wear.
   D. reduced pumping efficiency.
   E. any of the above.

_____ 37. A mud-gas separator –
   A. operates by applying vacuum to gas-cut well fluid.
   B. may operate under moderate back-pressure.
   C. prevents uncontrolled expansion of gas in outflowing mud.
   D. removes all entrained gas from mud.

_____ 38. The vacuum degasser –
   A. works better than the mud-gas separator on entrained gas.
   B. is used during a well blowout.
   C. burns gases as they enter the reserve tank.
   D. does none of the above.

Look again at the sections on **Duplex Mud Pumps** and **Triplex Mud Pumps**. Then answer the following questions.

_____ 39. A duplex mud pump is usually –
   A. centrifugal.
   B. triplex.
   C. double-acting.
   D. reciprocating.
   E. none of the above.

_____ 40. Fluid knock is caused by –
   A. large pieces of rock in the mud pump.
   B. preignition.
   C. loose pistons.
   D. gas.
   E. none of the above.

_____ 41. If a pump with a piston displacement of 300 cubic inches pumps 200 cubic inches of mud per stroke, its volumetric efficiency is 90 percent.
   A. True
   B. False

_____ 42. The most frequent cause of reduced volumetric efficiency is –
   A. leakage between piston and liner.
   B. gas-cut or air-cut mud.
   C. leakage between liner and pump body.
   D. leakage between valve and seat.
   E. none of the above.

_____ 43. Replacing a large liner in a reciprocating pump with a smaller liner _____ output pressure and _____ output volume.
   A. lowers; raises
   B. lowers; lowers
   C. raises; raises
   D. raises; lowers

_____ 44. Study table 3.3 At 60 spm, a duplex double-acting mud pump with 5½-inch liners and a 14-inch stroke pumps _____ gallons per minute.
A. 531
B. 388
C. 316
D. 257

_____ 45. Running at 50 strokes per minute, the pump in the above problem will pump about _____ gallons per minute.
A. 212
B. 263
C. 379
D. 428

_____ 46. A larger opening between suction line and dampener reduces hydraulic hammer more effectively than a smaller opening.
A. True
B. False

_____ 47. A desurger –
A. absorbs pressure variations.
B. removes gas-cut mud.
C. maximizes pump output.
D. reduces pump damage.

_____ 48. A single-acting triplex pump produces _____ pressure surges per crankshaft revolution.
A. one
B. two
C. three
D. four
E. six

_____ 49. A charging pump helps prevent –
A. piston knock.
B. shock load damage.
C. excessive pressure surges.
D. low volumetric efficiency.
E. all of the above.

_____ 50. A suction charging pump puts an additional work load on the reciprocating pump.
A. True
B. False.

The remaining questions are definitions you should know from Lesson 3. Match each definition on the right with a word or phrase on the left. Place the appropriate letter in the blank
next to the number.

_____ 51. Reserve pit

_____ 52. Barite

_____ 53. Mud-gas separator

_____ 54. Jet hopper

_____ 55. Desurger

_____ 56. Reactive solids

_____ 57. Duck's nest

_____ 58. Spm

_____ 59. Vortex finder

_____ 60. Duplex

_____ 61. Volumetric efficiency

_____ 62. Shale shaker

_____ 63. Nonreactive solids

_____ 64. Decanting mud centrifuge

_____ 65. Desilter

_____ 66. Vacuum degasser

_____ 67. Bentonite

_____ 68. Liner

_____ 69. Double-acting

_____ 70. Slug tank

_____ 71. Attapulgite

_____ 72. Triplex

_____ 73. Charging pump

_____ 74. Single-acting

_____ 75. Underflow

A. salt gel
B. removable inner cylinder surface
C. removes large cuttings from mud return
D. used to maintain positive pressure in
   suction line
E. used to separate unpressurized gas from
   mud.
F. mud components which do not adsorb
   water
G. separates solids under 2 and over 5
   microns
H. type of pump with two cylinders
I. type of pump with two pressure pulses
   per piston stroke
J. used to add dry mud materials to fluid
   mud
K. strokes per minute
L. mud components with an equivalent
   diameter of less than 2 microns
M. heavy material used to weight mud
N. depository for waste fluid and cuttings
O. used to remove high-pressure gas from
   drilling fluid
P. type of pump with three cylinders
Q. where heavy mud is mixed
R. removes both sand and silt from mud
S. liquid discharge end of a cone-shaped
   centrifuge
T. a low-gravity reactive solid
U. suction pulsation dampener
V. having one pressure pulse per piston
   stroke
W. partition of a reserve pit containing
   good mud
X. volume of fluid pumped divided by
   volume swept by pump pistons
Y. lower discharge stream from cone-
   shaped centrifuge

Lesson 4
THE DRILL STEM

Introduction and Early History

Drill Pipe

Tool Joints

Drill Collars

Drill Stem Auxiliaries

Operations Involving the Drill Stem

Lesson 4
THE DRILL STEM

INTRODUCTION AND EARLY HISTORY

The *drill stem* includes all items used for rotary drilling from the swivel to the bit, including the kelly, drill pipe, tool joints, drill collars, stabilizers, and miscellaneous other items of equipment (drill stem subs, reamers, and shock subs). The *drill string* is composed only of the drill pipe with attached tool joints that transmits fluid and rotational power from the kelly to the drill collars and bit. (Although by definition the kelly is a part of the drill stem, it is not discussed in this lesson.)

During the early days of rotary drilling, pipe, couplings, and drill collars of any given manufacturer often would not match similar products made by another company. Wall thicknesses differed, so inside and outside diameters of items used in the drill stem varied considerably. Additionally, threads would not always mate. This confusion led the American Petroleum Institute (API) to standardize threads and fittings. Specifications were also made for the types of material to be used, the methods of manufacturing, and the dimensions of pipe, threads, and mating connections.

Drill pipe thread failures were also a concern to operators and drillers during this early phase of drilling. Tool joints were first welded to drill pipe in 1934, and shrink-on tool joints were developed the following year. These innovations strengthened the drill stem.

An additional factor was responsible for the improvement in drill stem strength. Prior to the 1930s, drill pipe was run *in compression* (squeezed) to put weight on the bit. The use of drill collars to provide weight on the bit changed drilling techniques. Drill pipe now can be run *in tension* (stretched). Only the drill collars and not the drill pipe are allowed by the hoist to put weight on the bit. Running drill pipe in tension has improved straight-hole drilling techniques and reduced thread breaks and other failures.

DRILL PIPE

Drill pipe is the steel or aluminum tube used to transmit rotational power and drilling fluid to the bit at the bottom of the hole. Almost all drill pipe, fittings, and other rig equipment are made according to API specifications. Equipment that is not API-recommended must usually be specially ordered.

Design

Drill pipe and other tubular equipment are measured by the nominal outside diameter (OD) of the tube. The OD must be a specific measurement so that threaded fittings and pipe-handling tools fit properly. Although the OD of drill pipe, casing, tubing, and line pipe is the same for each size, the inside diameter (ID) varies with the weight per foot of length of pipe. Smaller ID and larger OD measurements improve torsional strength for deep drilling.

The standard length for drill pipe is categorized in three ranges. Range 1 (18 to 22 feet) is obsolete. Range 2 is the most widely used, and includes pipe 27 to 30 feet long. Range 3 is drill pipe of 38 to 45 feet. These lengths do not include the length of the tool joint of each end.

The four standard grades and strengths of seamless steel drill pipe are shown in table 4.1. (Grade D is no longer a standard inventory item.) *Minimum yield strength* refers to the force needed to stretch or compress the drill pipe until it is permanently distorted. *Minimum tensile strength* refers to the force necessary to pull the pipe apart.

TABLE 4.1
SEAMLESS STEEL DRILL PIPE

API Grade	Yield Strength (minimum psi)	Yield Strength (maximum psi)	Tensile Strength (minimum psi)
E	75,000	105,000	100,000
95(X)	95,000	125,000	105,000
105(G)	105,000	135,000	115,000
135(S)	135,000	165,000	145,000

Seamless drill pipe is *upset,* which means that the pipe is thicker at the ends for added strength. The three types of upset are internal upset (IU), external upset (EU), and internal-external upset (IEU). Upsets are necessary on drill pipe to provide safety in the weld area where the tool joint with its threaded connections is attached (fig. 4.1).

Dimensions and weights for grade E drill pipe are shown in fig. 4.2, from API Spec 5A. Nominal weights given in column

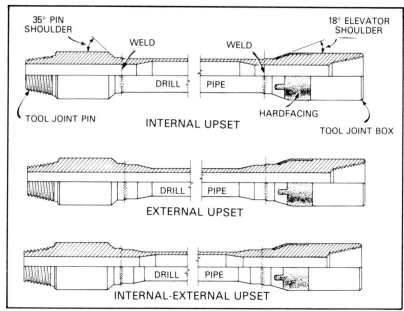

Figure 4.1. Upset

TABLE 6.8
UPSET DRILL PIPE FOR WELD-ON TOOL JOINTS
Dimensions and Weights
(Grade E)
See Fig. 6.8
See Appendix A for Metric Tables

1	2	3	4	5	6	7	8	9	10	11	12		13
				Calculated Weight		[2]Outside Diameter, $+\frac{1}{8}$, $-\frac{1}{32}$	[3]Inside Diameter at End of Pipe, $\pm\frac{1}{16}$	Length of Internal Upset $+1\frac{1}{2}$, $-\frac{1}{2}$	Length of Internal Taper, min.	Length of External Upset, min.	Length of External Taper,		Length End of Pipe to Taper Fadeout, max.
Pipe Size: Outside Dia., in. D	[1]Nominal Wt.: lb/ft	Wall Thickness, in. t	Inside Diameter, in. d	Plain End lb/ft w_{pe}	[4]Upset lb e_w	D_{ou}	d_{ou}	L_{iu}	m_{iu}	L_{eu}	min. m_{eu}	max.	$L_{eu}+m_{eu}$
INTERNAL-UPSET DRILL PIPE													
2⅞	10.40	0.362	2.151	9.72	3.20	2.875	1 5/16	1¾	1½	...	...	...	....
3½	9.50	0.254	2.992	8.81	4.40	3.500	2¼	1¾	...	...	...	...	....
3½	13.30	0.368	2.764	12.31	4.40	3.500	1 15/16	1¾	1½	...	...	...	....
3½	15.50	0.449	2.602	14.63	3.40	3.500	1 15/16	1¾	1½	...	...	...	....
*4	11.85	0.262	3.476	10.46	4.20	4.000	2 15/16	1¾	...	...	...	...	....
4	14.00	0.330	3.340	12.93	4.60	4.000	2¾	1¾	2	...	...	...	....
*4½	13.75	0.271	3.958	12.24	5.20	4.500	3⅜	1¾	...	...	...	...	....
*5	16.25	0.296	4.408	14.87	6.60	5.000	3¾	1¾	...	...	...	...	....
EXTERNAL-UPSET DRILL PIPE													
2⅜	6.65	0.280	1.815	6.26	1.80	2.656	1.815	...	...	1½	1½	...	4
2⅞	10.40	0.362	2.151	9.72	2.40	3.219	2.151	...	...	1½	1½	...	4
3½	9.50	0.254	2.992	8.81	2.60	3.824	2.992	...	...	1½	1½	...	4
3½	13.30	0.368	2.764	12.31	4.00	3.824	2.602	2¼	2	1½	1½	...	4
3½	15.50	0.449	2.602	14.63	2.80	3.824	2.602	...	...	1½	1½	...	4
*4	11.85	0.262	3.476	10.46	5.00	4.500	3.476	...	...	1½	1½	...	4
4	14.00	0.330	3.340	12.93	5.00	4.500	3.340	...	...	1½	1½	...	4
*4½	13.75	0.271	3.958	12.24	5.60	5.000	3.958	...	...	1½	1½	...	4
4½	16.60	0.337	3.826	14.98	5.60	5.000	3.826	...	...	1½	1½	...	4
4½	20.00	0.430	3.640	18.69	5.60	5.000	3.640	...	...	1½	1½	...	4
INTERNAL-EXTERNAL-UPSET DRILL PIPE													
4½	16.60	0.337	3.826	14.98	8.10	4.656	3 5/32	2½	..	1½	1	1½	....
4½	20.00	0.430	3.640	18.69	8.60	4.781	3	2¼	2	1½	1	1½	....
5	19.50	0.362	4.276	17.93	8.60	5.188	3 11/16	2¼	2	1½	1	1½	....
5	25.60	0.500	4.000	24.03	7.80	5.188	3 7/16	2¼	2	1½	1	1½	....
5½	21.90	0.361	4.778	19.81	10.60	5.563	4	2¼	2	1½	1	1½	....
5½	24.70	0.415	4.670	22.54	9.00	5.563	4	2¼	2	1½	1	1½	....

[1]Nominal weights (Col. 2), are shown for the purpose of identification in ordering.

[2]The ends of internal-upset drill pipe shall not be smaller in outside diameter than the values shown in Col. 7, including the minus tolerance. They may be furnished with slight external upset, within the tolerance specified.

[3]Maximum taper on inside diameter of internal upset and internal-external upset is ¼ in. per ft. on diameter.

[4]Weight gain or loss due to end finishing. See Par. 6.5.

[5]The specified upset dimensions do not necessarily agree with the bore and OD dimensions of finished weld-on assemblies. Upset dimensions were chosen to accommodate the various bores of tool joints and to maintain a satisfactory cross section in the weld zone after final machining of the assembly.

[6]By agreement between purchaser and manufacturer, the length of upset for Grade E drill pipe may be the same as for the higher grades listed in Spec 5AX.

*These sizes and weights are tentative.

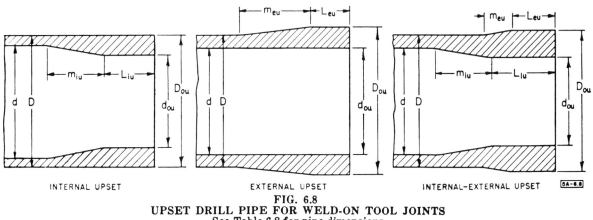

FIG. 6.8
UPSET DRILL PIPE FOR WELD-ON TOOL JOINTS
See Table 6.8 for pipe dimensions

Figure 4.2. API specifications for grade E weld-on tool joints (*Courtesy of API Spec 5A*)

2 are for the drill pipe only and do not include tool joint weight. Upsets for various sizes and weights of drill pipe are shown. High-strength drill pipe requires heavier and longer upsets than those used on grade E.

Figures 4.3 and 4.4 show the torsional, tensile, collapse, and internal pressure data for new drill pipe by size, weight, and grade.

Torsional Stength

Torsional strength concerns the degree to which pipe can be twisted. The torsional yield strength in foot-pounds for new drill pipe is given in fig. 4.3, from API RP 7G. For example,

TABLE 2.2
NEW DRILL PIPE TORSIONAL
AND TENSILE DATA

1	2	3	4	5	6	7	8	9	10
Size OD In.	Nom. Weight Thds & Couplings lb.	Torsional Data* Torsional Yield Strength, ft-lb				Tensile Data Based on Minimum Values** Load at the Minimum Yield Strength, lb;			
		E	95	105	135	E	95	105	135
2⅜	4.85	4760	6030	6670	8570	97820	123900	136940	176070
	6.65	6250	7920	8750	11250	138220	175080	193500	248790
2⅞	6.85	8080	10240	11320	14550	135900	172140	190260	244620
	10.40	11550	14640	16180	20800	214340	271500	300080	385820
3½	9.50	14150	17920	19800	25460	194270	246070	271970	349680
	13.30	18550	23500	25970	33390	271570	343990	380190	488820
	15.50	21090	26710	29520	37950	322780	408850	451890	581000
4	11.85	19470	24670	27260	35050	230750	292290	323050	415350
	14.00	23290	29500	32600	41920	285360	361460	399500	513650
	15.70	25810	32690	36130	46460	324120	410550	453770	583420
4½	13.75	25910	32820	36270	46630	270030	342040	378040	486050
	16.60	30810	39020	43130	55450	330560	418700	462780	595000
	20.00	36900	46740	51660	66420	412360	522320	577300	742240
5	16.25	35040	44390	49060	63080	328070	415560	459300	590530
	19.50	41170	52140	57600	74100	395600	501090	553830	712070
	25.60	52260	66190	73160	94060	530150	671520	742200	954260
5½	19.20	44070	55830	61700	79330	372180	471430	521050	669920
	21.90	50710	64230	70990	91280	437120	553680	611960	786810
	24.70	56570	71660	79200	101830	497220	629810	696110	895000
6⅝	25.20	70580	89400	98810	127050	489470	619990	685250	881040

* Based on the shear strength equal to 57.7% of minimum yield strength and nominal wall thickness. Minimum torsional yield strength calculated from Formula A.14, Par. A.9, Appendix A, rounded to nearest 10 ft-lb.

**Minimum tensile strength calculated from Formula A.12, Par. A.7, Appendix A, rounded to nearest 10 lb.

Figure 4.3. API torsional and tensile data for new drill pipe (*Courtesy of API RP 7G*)

4½-inch, 16.6 pound, grade E pipe in new condition has a torsional yield strength (resistance to failure due to the application of twisting force) of 30,810 foot-pounds. In other words, the pipe will not fail as a result of a twisting strain equivalent to a line pull of about 7,702 pounds on pipe tongs having a lever arm of 4 feet (7,702 pounds × 4 feet = 30,808 foot-pounds).

The torsional strength of a *tool joint* depends on several factors, including the strength of the steel, size, thread form, and friction of the mating surfaces of the connection. In general, the torsional strength of drill pipe is nearly always less than the torsional strength of tool joints.

Tensile Strength

Tensile strength concerns how much weight can be put on pipe. Tensile strength of drill pipe, based on minimum values (fig. 4.3) can be used to determine useful working depth for a particular grade of pipe. High-grade, greater-strength pipe must be used to reach greater depth. More weight per foot

---

TABLE 2.3
NEW DRILL PIPE COLLAPSE AND INTERNAL PRESSURE DATA

1	2	3	4	5	6	7	8	9	10
Size OD in.	Nom. Weight Thds & Couplings lb.	Collapse Pressure Based On Minimum Values, psi.				Internal Pressure At Minimum Yield Strength, psi.			
		E	95	105	135	E	95	105	135
2⅜	† 4.85	11040	13980	15460	19070	10500	13300	14700	18900
	6.65	15600	19760	21840	28080	15470	19600	21660	27850
2⅞	† 6.85	10470	12930	14010	17060	9910	12550	13870	17830
	10.40	16510	20910	23110	29720	16530	20930	23140	29750
3½	9.50	10040	12060	13050	15780	9520	12070	13340	17150
	13.30	14110	17880	19760	25400	13800	17480	19320	24840
	15.50	16770	21250	23480	30190	16840	21330	23570	30310
4	†11.85	8410	9960	10700	12650	8600	10890	12040	15480
	14.00	11350	14380	15900	20170	10830	13720	15160	19490
	†15.70	12900	16340	18050	23210	12470	15790	17460	22440
4½	13.75	7200	8400	8950	10310	7900	10010	11070	14230
	16.60	10390	12750	13820	16800	9830	12450	13760	17690
	20.00	12960	16420	18150	23330	12540	15890	17560	22580
5	†16.25	6970	8090	8610	9860	7770	9840	10880	13990
	19.50	10000	12010	12990	15700	9500	12040	13300	17110
	25.60	13500	17100	18900	24300	13120	16620	18380	23620
5½	†19.20	6070	6930	7300	8120	7250	9190	10160	13060
	21.90	8440	10000	10740	12710	8610	10910	12060	15510
	24.70	10460	12920	14000	17050	9900	12540	13860	17830
6⅝	25.20	4810	5310	5490	6040	6540	8280	9150	11770

NOTE: Calculations are based on formulas in Appendix A.
†These sizes and weights of drill pipe are not included in the drill pipe lists in API Spec 5A and 5AX.

Figure 4.4. API collapse and internal-pressure data for new drill pipe (*Courtesy of API RP 7G*)

means more square inches of metal are in tension because of the thicker wall. Also, more total weight is involved in a long drill string. Higher-grade material has greater tensile strength for each square inch of cross-sectional area. Of course, varying safety factors are employed at the discretion of the operator, who takes into account the physical condition of the pipe.

Look at fig. 4.3. Discounting any safety factor that allows for pulling more than the weight of the string (overpull) and with no drill collars, grade E, 16.6-pound pipe could be run to 19,913 feet (330,560 ÷ 16.6 pounds per foot) before it would pull apart because of its own weight if hanging in air. (The tensile minimum load figure given in the chart is divided by the nominal weight to find proper depth.) Allowance for overpull and for the weight of the drill collars must, however, be calculated to obtain a practical working depth for any drill string.

Using the tensile strength value of 330,560 pounds for the pipe discussed in the example, the following working depth can be calculated for 4½-inch, grade E, 16.6-pound pipe.

Tensile load at minimum yield	=	330,560 pounds
Overpull safety factor		− 50,000 pounds
Allowed working load	=	280,560 pounds
Drill collar weight		− 40,000 pounds
Practical working load limit, in air	=	240,560 pounds.

Again, the new working load limit is divided by the pipe's nominal weight. (240,560 ÷ 16.6 pounds per foot = 14,491 foot depth in air, plus the length of drill collars.) This figure is for new pipe; the figure would be reduced for good used pipe. Based on similar assumptions, the following approximate working depths apply for higher grades of drill pipe.

Grade E	15,000 feet
Grade 95 (X)	20,000 feet
Grade 105 (G)	25,000 feet
Grade 135 (S)	30,000 feet

In practice, combinations of sizes, weights, and grades of pipe are used to obtain the desired safety factor in tension. Usually, higher-grade and heavier-weight pipe is placed only in the upper portions of the drill string. For deep holes, 20,000 feet or more, several weights and grades are employed. Large-diameter pipe is sometimes used in the top of the drill string to obtain the needed strength, but doing so may require the use of extra pipe rams in the blowout preventers.

Collapse Resistance

Collapse resistance concerns how much external pressure can be put on pipe. Collapse resistance of drill pipe is an important consideration when making a *drill stem test* (DST). A DST is a test run before well completion in order to predict what a formation will produce. The testing tool may be run on the bottom of the drill string with no fluid inside the pipe above it. It is lowered into a wellbore full of drilling fluid. External pressure exists due to the hydrostatic pressure of the fluid outside the pipe. If the DST is run in a well containing 10-ppg fluid, a hydrostatic pressure gradient of 0.52 psi per foot of depth is present. If the DST is run with empty drill pipe above the testing tool, the result can be a collapsing pressure on the drill string as great as 5,200 psi at 10,000 feet. (See in Lesson 1 the section on formation pressures for an understanding of why the depth of the hole is multiplied by the pressure gradient of the mud in order to calculate the mud's pressure in pounds per square inch.) This external pressure tends to flatten the pipe unless a leak in the pipe or at a tool joint is present. Under such conditions a small hole may quickly wash out and cause a large leak, which would invalidate a DST made to obtain a pure sample of formation fluid.

Collapse resistance of drill pipe is reduced if the pipe is in tension, often the case during the DST if it becomes necessary to pull on a stuck packer. If calculations indicate that collapse resistance of the drill string will be exceeded by pulling on the pipe, it may be desirable to add fluid (a water cushion) inside the pipe to relieve the collapsing force caused by the hydrostatic pressure outside the pipe.

Burst Strength

Burst strength concerns how much internal pressure can be put on pipe. Data for burst strength of new drill pipe is given in fig. 4.4. This API table shows that an internal pressure of 9,500 psi will burst 5-inch OD, 19.5-pound, Grade E pipe by exceeding the minimum yield strength of the material – the internal pressure that causes a pipe in new condition to rupture. Tabulations are available for used pipe. Class 3 used pipe of the same weight and grade, for example, bursts at less than 6,000 psi. Such high pressures do not usually occur except when making a DST of high-pressure gas zone (when the gas is traveling up the drill pipe) or when pumping on a fracture treatment. When exceptionally high internal pressures are expected, careful review must be made of engineering data, safety factors, and mechanical condition of the pipe.

Failures

Fatigue break is the most common drill pipe failure, often taking place in the slip area as a result of deep scratches or metal tears caused by the pipe turning in the slips. A fatigue break may appear as a hole and be called a *washout*. A square or spiral break may be called a *twistoff*. Both terms are

Figure 4.5. Washout

Figure 4.6. Twistoff

misleading. A washout occurs when drilling fluid is forced through a small opening in the metal caused by fatigue. Fluid abrasion erodes the metal and enlarges and rounds off the edges of the hole (fig. 4.5). Modern drill pipe has torsional yield strength, and will wrap up rather than twist off if the bit becomes stuck and rotation is continued. The twistoff usually seen in the field is caused by a fatigue crack that extends around the pipe (fig. 4.6). During the last part of the failure, the pipe wall was torn and slightly twisted.

Three types of fatigue failure are generally found:

(1) *pure fatigue* – a break without any previous visible cause;
(2) *notch fatigue* – a break associated with a mechanical notch; and
(3) *corrosion fatigue* – a break where the notch is caused by corrosion.

Pure fatigue. Metal is weaker under working loadings than under static conditions. Steel is capable of absorbing *dynamic loading,* or cycles of stress, for an infinite number of cycles if the stress is kept under a certain limit. The chemical composition, surface finish, and tensile properties of steel in part determine drill pipe fatigue limit. *Pure fatigue* is a metal break that has no visible cause.

Metal fatigue in drill pipe cannot actually be measured. The fatigue strength of drill pipe steel is said to be approximately one-half its tensile strength. Calculations can be made for a given set of downhole conditions to indicate the fatigue percentage that may be expected. The best indicator, however, is the actual frequency of failures.

Drill pipe is subjected to cyclic stresses in tension, compression, torsion, and bending. Tension and bending are the most critical. The major factor in drill pipe fatigue is cyclic bending when pipe is rotated in a hole that has a change in direction (a dogleg). Nevertheless, fatigue can occur in straight-hole drilling as well as in directional drilling. It can occur in spite of the fact that drill stem weight is maintained to keep the drill string in tension and no crooked pipe occurs in the drill string. When pipe is rotated through a dogleg in the hole, each side of the pipe goes through cycles of stress (from tension to compression) with each rotation. Drill pipe rotated at 100 rpm makes 144,000 revolutions per day if run continuously. In seven days, more than a million stress cycles on the pipe occur. If the stress exceeds the fatigue strength of the metal, the pipe fails.

Notch fatigue. Surface imperfections, which may be either mechanical notches or metallurgical defects in the steel itself, greatly affect the fatigue limit. Aside from the initial distortion of the steel grain structure, a notch raises the stress level and speeds the breakdown of the metal structure. Notches and pits are therefore called *stress risers* or *stress concentrators.* If a notch is on a portion of drill pipe not subject to stress, it has little effect. If, however, a notch occurs in the *area of maximum bending,* within 20 inches of a tool joint, it can start a fatigue break. A longitudinal notch is relatively harmless, but a transverse notch leads to failure. An extensive notch with a rounded bottom may not fail immediately. A small scratch with a sharp, V-shaped bottom acts as a stress riser and quickly causes failure. Some steel is more sensitive to notches than other steel. For example, hard, brittle steel fails more quickly than ductile steel. Various surface dents and scratches that can cause drill pipe notch failure are –

(1) slip marks, cuts, and scratches;
(2) tong marks;
(3) spinning chain marks and scratches;
(4) stencil markings;
(5) grooves caused by rubber protectors;
(6) electric arc burns; and
(7) downhole notching by formation and junk cuts.

Tong marks are probably the worst-looking defects produced on drill pipe in the field. Tong marks are long, deep, and frequently sharp notches. Because such marks are longitudinal, they seldom lead to notch failure. Nevertheless, with even slight deviation from vertical, they can become stress concentration points. The application of the tongs to the body of the drill pipe instead of to the tool joint is a bad practice because of the possibility of crushing the pipe.

Both tongs should always be used when making up or breaking out drill pipe. If only one set of tongs is used, the pipe may turn in the slips (wedge-shaped pieces of metal that support the drill stem in the rotary table). A deep scar may result. Such a scar is usually circumferential, but may be spiral if the pipe drops while slipping.

The tool joint should be as close to the rotary as possible during makeup and breakout. If a tool joint is positioned more than maximum height above the rotary slips, the pipe does not have enough strength to resist bending. This height depends on the makeup torque, the length of the tong handles, the angle between the tong handles, and the yield strength of the pipe. API RP 7G shows that the maximum height for $4\frac{1}{2}$-inch, 16.6-pound, grade E drill pipe with $6\frac{1}{4}$-inch $\times$ $3\frac{1}{4}$-inch tool joints and tongs at a 90-degree angle is 3.4 feet above the

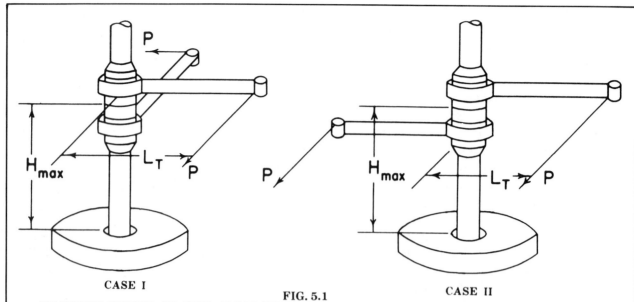

CASE I   FIG. 5.1   CASE II

MAXIMUM HEIGHT OF TOOL JOINT ABOVE SLIPS TO PREVENT BENDING DURING TONGING

5.8 Drill Pipe Bending Resulting From Tonging Operations. It is generally known that the tool joint on a length of drill pipe should be kept as close to the rotary slips as possible during make-up and break-out operations to prevent bending of the pipe.

There is a maximum height that the tool joint may be positioned above the rotary slips and the pipe resist bending, while the maximum recommended make-up or break-out torque is applied to the tool joint.

Many factors govern this height limitation. Several of these which should be taken into most serious consideration are:

(1) The angle of separation between the make-up and break-out tongs, illustrated by Case I and Case II, Fig. 5.1. Case I indicates tongs at 90° and Case II indicates tongs at 180°.

(2) The minimum yield strength of the pipe.

(3) The length of the tong handle.

(4) The maximum recommended make-up torque.

$$H_{max} = \frac{.053\ Y_m\ L_T\ (I/C)}{T} \quad (Case\ I).............5.81$$

$$H_{max} = \frac{.038\ Y_m\ L_T\ (I/C)}{T} \quad (Case\ II)...........5.82$$

Where:

H_{max} = Height of tool joint shoulder above slips—ft
Y_m = Minimum tensile yield stress of pipe—psi
L_T = Tong arm length—ft
P = Line pull (Load)—lbs
T = Make-up torque applied to tool joint ($P_x L_T$)—lb ft
I/C = Section Modulus of pipe—in.³

Constants .053 and 0.038 include a factor of 0.9 to reduce Y_m to proportional limit. (See Par. 5.3)

Sample Calculation:

Assume: 4½ in., 16.60 lb/ft, Grade E drill pipe, with 4½ in. X.H. 6¼ in. OD, 3¼ in. ID tool joints.

         Tong arm 3½ ft

         Tongs at 90° (Case I)

Using equation 5.81:

$$H_{max} = \frac{.053\ (Y_m)\ (I/C)\ (L_T)}{T}$$

Y_m = 75,000 psi (for Grade E)

I/C = 4.27 in.³ (Table 5.1)

L_T = 3.5 ft

T = 17,000 ft-lb (from Table 2.12)

$$H_{max} = \frac{.053\ (75,000)\ (4.27)\ (3.5)}{17,000} = 3.4\ ft$$

TABLE 5.1
SECTION MODULUS VALUES

1	2	3
Pipe O.D. in.	Pipe Weight Nominal lbs/ft	$\dfrac{I}{C}$ cu. in.
2⅜	4.85	0.66
	6.65	0.87
2⅞	6.85	1.12
	10.40	1.60
3½	9.50	1.96
	13.30	2.57
	15.50	2.92
4	11.85	2.70
	14.00	3.22
	15.70	3.58
4½	13.75	3.59
	16.60	4.27
	20.00	5.17
	22.82	5.68
	24.66	6.03
	25.50	6.19
5	16.25	4.86
	19.50	5.71
	25.60	7.25
5½	19.20	6.11
	21.90	7.03
	24.70	7.84
6⅝	25.20	9.79

Figure 4.7. Calculation of maximum height of tool joint above slips (*Courtesy of API RP 7G*)

rotary. See Case I in the first column of figure 4.7. Case II, where the tongs are 180 degrees apart, requires that the tool joint be a shorter distance above the rotary in order not to exceed the yield strength of the steel and bend the pipe. In figure 4.8, drill pipe is shown which has been bent by using only one set of tongs or by setting the slips too low, leaving the tool joint too high above the rotary.

Another damaging effect of using only one set of tongs when going in the hole is that the driller, realizing that the pipe will turn in the slips if the proper makeup torque is applied, applies less torque. The result may be a loose connection, galled shoulders, and quite often, last-engaged thread failures. Another expensive problem caused by low makeup torque is excessive downhole tightening of the joint during drilling. The resulting hard-to-break connection can mean deep tong marks, and in an environment of sharp sand the connection may produce extensive casing wear and cause high rates of tool jont OD wear.

Rotary table slips have fine serrations that ordinarily do not leave injurious marks on drill pipe. However, if the slips are mistreated, worn, or carelessly handled, they can score the pipe, usually in the transverse or horizontal direction. The practice of rotating drill pipe with the slips can leave a dangerous transverse notch if the pipe turns in the slips. Slips with worn, mismatched, or incorrectly installed gripping elements can allow one or two teeth to catch the full load, causing a deep notch and potential failure. Damage is particularly likely when a combination of new and old components is used (fig.4.9).

Slips must not catch the drill pipe during the process of going in the hole. Stopping downward motion of drill pipe with the slips can cause necking down of the pipe in the slip area, as well as excessive loads on the rotary. Slip-area damage, particularly crushing of drill pipe, can also be caused by the use of a damaged or worn rotary table, master bushing, or slip bowl.

Other poor operating practices, such as allowing the slips to ride the pipe in coming out of the hole, can cause damage to the slips. If the pipe does not hold in the rotary slips because of worn gripping elements, the elements should be replaced immediately. If pipe slides through the slips until a tool joint contacts the slips, the sudden stop may result in a bouncing of the string that kicks the slips out of the master bushing, and the pipe disappears down the hole.

Deep wells involve heavy loads on the full drill string. Slip marks on the pipe are extremely hazardous and may occur even though great care is taken in setting the slips. For this reason, some deep-well operators have eliminated the use of slips and instead use two sets of elevators.

Corrosion fatigue. Corrosion fatigue, or metal fatigue in a corrosive environment, is a common cause of drill stem failures. Corrosion can take many forms and may combine

Figure 4.8. Drill pipe bent from setting the slips too low or using only one tong

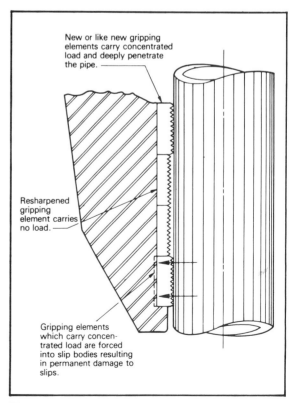

Figure 4.9. Results from using old and new gripping elements together in slips

109

with erosion, abrasive wear, and notch failures to cause severe damage. Hydrogen sulfide, carbon dioxide, or an acid mud in the wellbore can add to corrosion fatigue. Several forms of corrosion may take place simultaneously. Corrosive agents that cause metal pitting are responsible for most corrosion fatigue failures.

TOOL JOINTS

Many different designs have been developed over the years for the threaded joints that connect lengths of drill pipe. However, practically all tool joints made today are flash-weld joints (fig. 4.10).

Tool joint nomenclature is given in figure 4.11. Note that *hardfacing* (an extremely hard material applied to the surfaces of drill string components) is an optional feature and may take several forms. Drill pipe is identified with a steel die stencil on the pin's taper (fig. 4.12). Placing the stencil on the box end may allow it to be scoured away by the mating pins or by abrasive drilling fluid.

Design

Tool joint dimensions have been established for standard weights and grades of drill pipe. In 1968 a new series of rotary-shouldered connections was adopted as standard by API. In this style of connection, the size designation is a two-digit number that refers to the dimension of the pin. This size

Figure 4.10. API standard welded tool joint

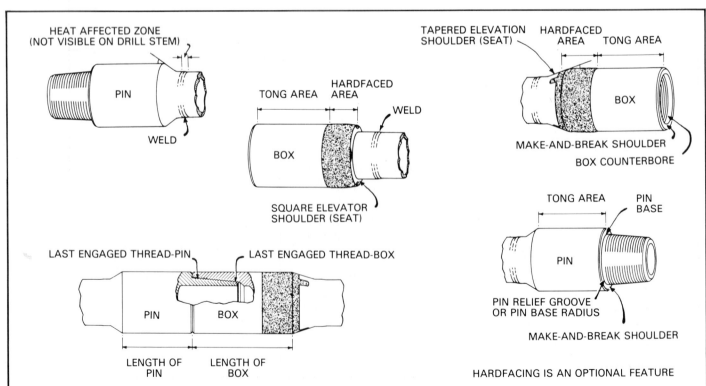

Figure 4.11. API tool joint nomenclature

Company symbol, month welded, year welded, pipe mill symbol and drill pipe grade code are to be stencilled at the base of the pin.

MONTH AND YEAR WELDED

Month	Year
1 Through 12	Last two digits of year

DRILL PIPE GRADE CODE

Grade	Symbol
N-80	N
E	E
C-75	C
X-95	X
G-105	G
S-135	S
V-150	V
Used	U

HEAVY WEIGHT DRILL PIPE
(Double Stencil Pipe Grade Code.)

PIPE MILL CODE

Pipe Mill	Symbol
Armco	A
J & L Steel	J
U. S. Steel	N
Wheeling — Pittsburgh	P
Youngstown	Y
Dalmine S.P.A., Italy	D
Falck, Italy	F
TAMSA	T
Nippon Kokan Kabushiki	K
Vallourec	V
Mannesmannrohren-Werke	M
Sumitomo Metal Ind.	S

SAMPLE MARKINGS AT BASE OF PIN

1	2	3	4	5
ZZ	6	70	N	E

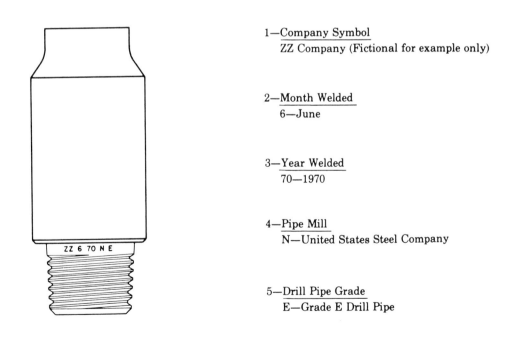

1—Company Symbol
ZZ Company (Fictional for example only)

2—Month Welded
6—June

3—Year Welded
70—1970

4—Pipe Mill
N—United States Steel Company

5—Drill Pipe Grade
E—Grade E Drill Pipe

Figure 4.12. Tool joint marking to identify drill pipe
(*Courtesy of API RP 7G*)

designation is called NC, or *numbered connection.* The old IF (internal flush) and FH (full hole) connections are becoming obsolete, although still used to some extent. Equivalent NC connections are available for most of the old IF and some of the FH tool joints. There are no NC equivalents for API Regular, which is still the standard connection used on swivel subs. The only API connection used on 5½-inch drill pipe is 5½-inch FH.

Care and Handling

Pin and box threads and shoulders of tool joints should be thoroughly cleaned before being added to the drill string. Cleaning removes foreign material and permits proper makeup, permits better inspection, and increases the life of connections by eliminating abrasive materials. Damaged connections should never be run in the hole.

Thread protectors made of pressed steel prevent most of the tool joint damage that can occur in moving and racking. Protectors are also available in plastic or rubber. During picking up or laying down drill pipe, threads and shoulders of tool joints should not be permitted to strike steel on the pipe walk or ramp. When coming out of the hole, elevators must not strike the tool joint box shoulder. When going into the hole, the tool joint pin should not strike the shoulder of the tool joint box. Poor stabbing can cause damage.

Tool joints are subjected to high torsional loads in service. *Galling* (damage to steel surfaces caused by friction) may occur on the threads and shoulders unless there is a separating film of lubricant. Thread compounds provide this film, and the appropriate compound will ensure proper makeup if the correct torque is applied. Ordinary greases and oils do not have sufficient body to provide good lubrication under a high torsional load. Thread compounds containing soft metallic fillers do not squeeze out readily, and they can withstand high stress without breaking down. Compounds containing 40 percent to 60 percent by weight of finely powdered metallic zinc are recommended for rotary-shouldered joints. Compounds containing 60 percent or more metallic lead are used on drill collar connections. Thread compounds should not be thinned for ease of application in cold weather. Dilution reduces the filler metal content, thereby decreasing the effectiveness of the compound.

Figure 4.13 is an illustration of how tool joints should be lubricated. It is most important that the last swipe of the brush spread thread compound on the shoulder face. Thread lubricants compounded for tubing or casing should not be used on tool joints. They are slick and will allow the joint to be overtightened with a low torque application. Several instances have occurred in which tool joints have been severely damaged by excessive makeup because these lubricants were used. The damage was mostly in the form of stretched or broken pins or swelled boxes.

Figure 4.13. Tool joint lubrication with proper zinc compound

Wobble

A tool joint is said to wobble if any movement occurs between the surfaces of the mating box and pin. This condition can be caused by improperly mated shoulders, low makeup torque, or other mistakes. If the joint is allowed to wobble for even a short time, the threads and shoulders of both box and pin are damaged. If the wobble is allowed to continue, failure and a subsequent fishing job result.

Wobble can occur under most drilling conditions if correct practices are not followed. Makeup tonging must be done correctly, especially in deep drilling, when the tensile load on tool joint connections is high. In crooked or directional holes, downhole torque and bending loads on tool joints are also high. Such conditions may permit the connection to tighten during drilling, resulting in hard-to-break joints, belled boxes, or stretched pins.

Ordinarily, the earliest indication of wobble is a dry or muddy appearance of pins. This condition shows that tool joints are not pressure-tight, but are allowing slight leakage of the thread compound, which is then displaced by drilling fluid. Often, close examination of such connections is required to reveal any damage. Obviously, a crowned shoulder can neither form an effective pressure seal against leakage nor provide support to prevent further wobble, regardless of how tightly the joint is tonged. Both the box joint and the pin joint should be removed from the string until they have been repaired. If damage is not too great to either threads or shoulders, field repair may be made to the shoulders of the damaged joints.

Close examination of dry pins and boxes may reveal sharp, crested threads that are lapped and worn. Such sharp threads may exist for some time before wobble causes more obvious damage, such as broken threads. When such a connection is backed out, the broken threads often become fouled (fig. 4.14). It may even be impossible to break the pin all the way out of the box. Often a *gall* (surface damage caused by friction) is seen on the shoulders of wobbled joints, and it is blamed for the trouble. Actually the gall may be the result of the wobble. In such cases, a simple wobble causes the shoulders to crown; then subsequent working of the joint and leakage wash out the lubricant. Shoulder contact exists only between high spots; and when the connection is backed out, galling may occur between these overloaded, unlubricated high spots.

Several conditions may contribute to wobble. The first and most dangerous is the presence in the string of tool joints that have already wobbled. The shoulders of such damaged pins or boxes are almost invariably crowned and are incapable of furnishing sufficient support to prevent further wobble and damage. Wobble spreads throughout the string (fig. 4.15). Each damaged pin may wobble in several boxes, and each damaged box may wobble on several pins.

Figure 4.14. Fouled and broken pin threads

Figure 4.15. Damage transmitted from pin shoulder to box shoulder

113

Figure 4.16. Drill pipe and drill collars

DRILL COLLARS

A drill collar (fig. 4.16) is a heavy, thick-walled tube that is used between the drill pipe and the bit in order to–

(1) provide weight to the bit for drilling;
(2) maintain weight to hold the drill string in tension;
(3) provide the pendulum effect to cause the bit to drill a nearly vertical hole; and
(4) support and stabilize the bit in order to drill new hole aligned with hole previously drilled.

Weight on the Bit

The amount of weight required on the bit depends on the kind of formation being drilled, the diameter of the hole, the type of bit, the tendency of the hole to deviate from vertical, and other variables. Standard drill collars are 30 feet long, and the weight of a drill collar depends on the OD and the bore diameter. Outside diameter is as close to hole diameter as possible. Inside diameter is large enough to permit circulation with minimum pressure drop. In figure 4.17, drill collar standards with the numbered connections and standard OD and bore diameters are given.

Figure 4.17. API specifications for drill collars (*Courtesy of API Spec 7*)

TABLE 6.1
DRILL COLLARS
All dimensions in inches, unless otherwise specified.
Collars NC23-31 and NC77-110 are tentative. See Special Note page 3. See Fig. 6.1.

1	2	3	4	5	6
Drill Collar Number*	Outside Dia, *D*	Bore, +1/16 −0 *d*	Length, ft, ±6 in. *L*	Bevel Dia, ±1/64 D_F	Bending Strength Ratio
NC23-31 (tentative)	*3 1/8*	*1 1/4*	*30*	*3*	*2.57:1*
NC26-35 (2 3/8 IF)	3 1/2	1 1/2	30	3 17/64	2.42:1
NC31-41 (2 7/8 IF)	4 1/8	2	30	3 61/64	2.43:1
NC35-47	4 3/4	2	30	4 33/64	2.58:1
NC38-50 (3 1/2 IF)	5	2 1/4	30	4 49/64	2.38:1
NC44-60	6	2 1/4	30 or 31	5 11/16	2.49:1
NC44-60	6	2 13/16	30 or 31	5 11/16	2.84:1
NC44-62	6 1/4	2 1/4	30 or 31	5 7/8	2.91:1
NC46-62 (4IF)	6 1/4	2 13/16	30 or 31	5 29/32	2.63:1
NC46-65 (4IF)	6 1/2	2 1/4	30 or 31	6 3/32	2.76:1
NC46-65 (4IF)	6 1/2	2 13/16	30 or 31	6 3/32	3.05:1
NC46-67 (4IF)	6 3/4	2 1/4	30 or 31	6 9/32	3.18:1
NC50-70 (4 1/2 IF)	7	2 1/4	30 or 31	6 31/64	2.54:1
NC50-70 (4 1/2 IF)	7	2 13/16	30 or 31	6 31/64	2.73:1
NC50-72 (4 1/2 IF)	7 1/4	2 13/16	30 or 31	6 43/64	3.12:1
NC56-77	7 3/4	2 13/16	30 or 31	7 19/64	2.70:1
NC56-80	8	2 13/16	30 or 31	7 31/64	3.02:1
6 5/8 REG	8 1/4	2 13/16	30 or 31	7 45/64	2.93:1
NC61-90	9	2 13/16	30 or 31	8 3/8	3.17:1
7 5/8 REG	9 1/2	3	30 or 31	8 13/16	2.81:1
NC70-97	9 3/4	3	30 or 31	9 5/32	2.57:1
NC70-100	10	3	30 or 31	9 11/32	2.81:1
NC77-110 (tentative)	*11*	*3*	*30 or 31*	*10 17/64*	*2.78:1*

*The drill collar number (Col. 1) consists of two parts separated by a hyphen. The first part is the connection number in the NC style. The second part, consisting of 2(or 3) digits, indicates the drill collar outside diameter in units and tenths of inches. The connections shown in parentheses in Col. 1 are not a part of the drill collar number; they indicate interchangeability of drill collars made with the standard (NC) connections as shown. If the connections shown in parentheses in column 1 are made with the V-0.038R thread form (as provided in Par. 9.4) the connections, and drill collars, are identical with those in the NC style. Drill collars with 8 1/4 and 9 1/2 inches outside diameters are shown with 6 5/8 and 7 5/8 REG connections, since there are no NC connections in the recommended bending strength ratio range.

Drill collars weigh less in mud than in air because of the buoyancy of the mud. The heavier the mud, the greater is the buoyancy effect and the lighter the apparent weight of the collars. Extra collars are usually employed to offset the buoyancy effect. Field performance shows that drill collar weight must exceed the weight that is needed on the bit during drilling, with no weight to be supplied by the drill pipe. The usual practice is to use 10 percent to 30 percent excess drill collar weight over the amount needed on the bit to prevent buckling of the drill pipe.

Drill collar weight to the bit may be as little as a few thousand pounds, or in excess of 100,000 pounds, depending on the bit and hole size. Outside diameters of standard drill collars may vary from 3⅛ inches to 11 inches. The weights of drill collars vary from 21 to 306 pounds per foot. As few as six collars or as many as forty-eight may be used.

A drilling weight of 30,000–60,000 pounds is common. To obtain that much useful weight in 10-ppg mud, which has a buoyancy factor of 0.847, about 30,000–70,000 pounds of collar weight is needed. Excess weight of 15 percent is common, making the total collar weight approximately 40,000–80,000 pounds. The 80,000 pounds can be obtained by using an assembly of eighteen 8-inch × 2¹³⁄₁₆-inch drill collars; fewer larger-diameter collars can be used to obtain the same weight. Resistance to bending is best provided by drill collars of the largest possible size that can be safely run into the hole. Square drill collars are much stiffer than round collars of equal size. A point to remember is that drill collars lose weight as they are eroded by wear and by the drilling fluid coming out of the hole.

Holding the Drill Pipe in Tension

Drill collar weight must be calculated to be heavy enough so that the drill pipe is never allowed to buckle (fig. 4.18). When drill pipe is subject to buckling, metal fatigue failures result. Furthermore, the body of the drill pipe wears rapidly near the center of the joint, and tool joints wear quickly due to abrasion on the wall of the hole. The drill collar weight should be adequate to supply the load on the bit, with enough weight over that amount to keep the drill pipe in tension. Then the string of drill pipe will remain relatively straight as it is rotated. Because the pipe does not bend as it would under a compressive load, fewer fatigue failures occur and wear of the pipe and tool joints is slower.

Providing Pendulum Effect

Pendulum effect may be defined as the tendency of the drill stem to hang in a vertical position, due to the force of gravity. The heavier the pendulum, the stronger is its tendency to remain vertical and the greater the force needed to cause the drill stem to deviate from vertical. If the drill stem is suspended in any position other than vertical, the force of gravity

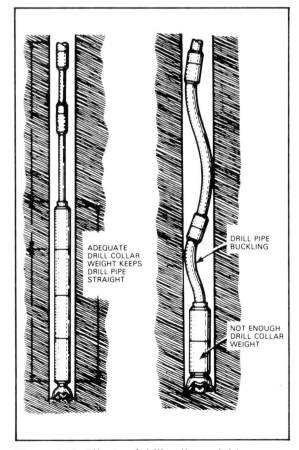

ADEQUATE DRILL COLLAR WEIGHT KEEPS DRILL PIPE STRAIGHT

DRILL PIPE BUCKLING

NOT ENOUGH DRILL COLLAR WEIGHT

Figure 4.18. Effects of drill collar weight

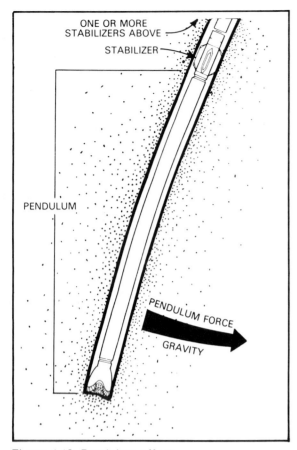

Figure 4.19. Pendulum effect

tends to pull the drill stem to vertical. If the hole deviates from vertical, the bit and drill collars tend to lie on the low side of the hole and seek to drill a path more nearly vertical.

The pendulum force works on the bottom of the drill string and tends to restore the drill string to vertical in an inclined hole (fig. 4.19). The length of the pendulum is that section of the drill collar string between the bit and the lowest point tangent to the side of the hole. This point should be as high as practical so that the pendulum will be as long as possible. The longer the pendulum, the greater is its tendency to seek a vertical position.

Weight of the drill collar is important in achieving the pendulum effect because a heavier pendulum is more effective than a lighter one. Stiffness of the drill collar assembly is also important because a stiff assembly has a higher point of tangency than a limber assembly. To achieve both weight and stiffness, large, heavy drill collars are recommended. Undersized and limber drill collars, having large clearance between the collar OD and the wall of the hole, make it harder to keep the hole vertical. Excessive weight applied to a limber drill collar string tends to bend or flex the collars near the bit, shortening the pendulum considerably.

Stabilizing the Bit

Large, square drill collars or stabilizers immediately above the bit stabilize the drill stem in the hole in spite of forces that tend to deviate the hole. Packed-hole assemblies using square drill collars or stabilizers guide the bit to drill a true extension of the previously drilled hole. The term *packed hole* refers to the fact that the square drill collars or stabilizers in the lower part of the assembly are only about ⅛ inch smaller in diameter than the hole.

Bit stabilization helps make straight hole and ensures proper bit performance because the bit is made to rotate on its axis. The bit is prevented from wobbling or walking on the bottom of the hole, and the cutting structure of the bit is uniformly loaded. An unrestrained bit may drill an oversize hole, produce unusual bit wear, and slow the rate of penetration. Bits drill faster and last longer when properly stabilized.

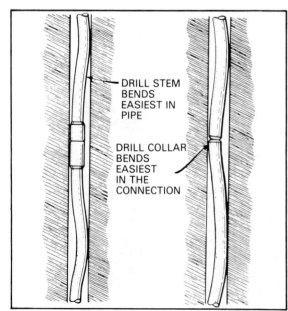

Figure 4.20. Bending points in drill pipe and drill collars

Failures

Drill collar connections are basically the same as the connections on drill pipe tool joints. Drill pipe failures take place in the pipe rather than in the tool joints, but drill collar failures usually take place in the threaded connections because of cyclic bending stresses. Fatigue failures occur at the point where bending is concentrated. The drill pipe wall is much thinner than the tool joint and is more easily bent. In drill collars, most of the bending must take place in the threaded connection (fig. 4.20). Since most of the bending takes place in the joint, most of the failures will occur there.

Drill collars can bend while rotating in the hole. A single 8-inch drill collar on the rack may appear to be so heavy and stiff that it could not bend, but when three are standing in the derrick, a definite bend can be seen. Twenty of these collars may be stacked in the hole, making a long and limber column that will bend when compressed.

In any bent member, the metal fibers on the outer side of the bend are under tension. Fibers on the inner side of the bend are under compression. If this member is rotated one-half turn, the fibers that were under tension are placed under compression. In every drill pipe and drill collar rotated in the hole, this continuous movement from tension to compression and back to tension takes place with each rotation of the drill stem (fig. 4.21). If the surface of any drill stem member is cracked, every time this crack comes to the tension side of the bend it tends to enlarge. Such cracks are called *fatigue cracks*.

Fatigue cracks originate at the areas where stress is concentrated–where bending stress is highest. Drill collars have two main areas of stress concentration–one in the pin, and one in the box. When the drill collar joint is properly made up and assembled, the shoulders support the pin so that it is rigidly held by the box. Then the weakest section is the narrow cross section in the bottom of the box, near the end of the pin. The other weak area is the first or second thread near the base of the pin. If the joints are not properly made up, the shoulders do not adequately support the pin in bending. If the shoulders are too narrow, they do not support the bending load imposed by the pin. In either case, severe bending stresses concentrate in the first and second threads, near the base of the pin. The result is a fatigue crack that starts between threads in this area of stress concentration. Boxes may also crack.

If there has been proper makeup control on the drill collar joint, fatigue failures may occur first in the bottom of the box. When makeup of the drill collar joint is poor, failures occur first at other points. As the collar above the joint bends back and forth, the pin bends back and forth. Too little compression in the drill collar joint shoulder does not keep it tight at all times, and the shoulder on the tension side opens up. The drill collar above the shoulder bends to the right; the shoulder opens up on the left and has additional compression on the right. A severe tension load is applied to the pin threads by the bending. As the drill collar rotates one-half turn, the side that was previously in compression is now in tension. The side that was in tension is now in compression. This rocking back and forth produces symptoms that can be recognized as a loose joint. Characteristically, with this situation, the pin is dry, and the area around the shoulder is a dull gray color. These indicators mean that fluid has been lapping in and out between the shoulders. Every time the shoulder opens up, fluid comes in; as the shoulder goes into compression, fluid is squirted out. This lapping of the drilling fluid removes some of the metal at the shoulders to produce the characteristic dull gray color.

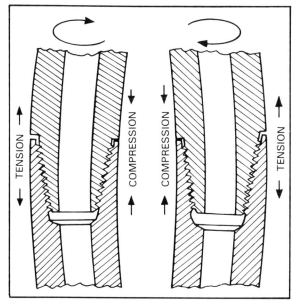

Figure 4.21. Tension and compression in a drill collar

117

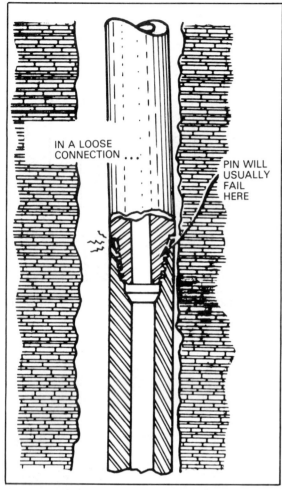

IN A LOOSE
CONNECTION . . .

PIN WILL
USUALLY
FAIL
HERE

Figure 4.22. Point of pin and failure on a loose drill collar connection

Tubing and casing threads have continuous seal in the threads themselves. That is, contact is continuous at all points along the thread profile. The threads themselves produce the pressure seal. In tool joint and drill collar threaded connections, the shoulder is the only seal. Shoulders must be kept smooth and clean. They must seal so tightly that even in bending they never separate or permit fluid to pass. If mud passes across the shoulder, three things happen:

(1) the lubricant between the shoulders washes out;
(2) the surface of the shoulder begins to lap; and
(3) if enough mud moves through, lubricant is washed away from the threads as well as from the shoulders.

In an operating joint that is loose (fig. 4.22), failures normally occur in the first or second thread near the base of the pin. Occasionally, failure occurs at the third or fourth thread. Such failures usually occur when the joint is relatively large compared with the size of the drill collar or when severe OD wear exists. In this case, the box shoulder is thin or narrow and so weak that it cannot support the bending stresses imposed by the pin. Its torsional strength is also low. The shoulder flares open as the pin bends, creating high bending stresses on the inside of the box near the mouth. This action starts fatigue of the box threads close to the shoulder. Such fatigue failures are frequent in areas where drillers use high rotary speeds and the drill collars have relatively large pins and thin shoulders. These conditions can also result in swelled, or belled, boxes—particularly in small-sized collars. Slick thread compounds aggravate this problem. Box failures of this type are rare if the joint has a good balance of pin size relative to the diameter of the box.

In addition to failures produced by fatigue, drill collar failures can result from using too much makeup torque. These failures occur in the first or second thread near the base of the pin. The pin and box members serve as a simple screw jack. This screw jack assembly is used to pull the two shoulders tightly together. The drilling rig has enough power to break something if pulling is continued after the shoulders are tight. The tongs or the pulling line breaks, the pin pulls, or the box swells. The pin actually screws through the box. Enough torque must be used, but too much torque causes failures.

API RP 7G lists recommended makeup torques for various sizes of drill collars and types of connections. A common error in determining the makeup torque for a given size and type of drill collar is to fail to take into account the wear of the drill collar and the actual diameter resulting. An example is an

NC-50 drill collar that is 7 inches (OD) × $2^{13}/_{16}$ inches (bore). It requires a minimum of 32,000 foot-pounds of makeup torque. However, if it is worn down to a 6¼-inch OD, the makeup torque should be reduced to a minimum of 22,800 foot-pounds. Another common error is to use tool joint specifications instead of drill collar specifications. An API NC-46 tool joint that is 6 inches (OD) × 3¼ inches (bore) requires a maximum of 18,000 foot-pounds. The same thread used on a 6-inch × $2^{13}/_{16}$-inch drill collar requires a minimum of 22,200 foot-pounds of makeup torque.

Handling

Drill collars are heavy and awkward to pick up, make up, and handle in the derrick. They must be moved carefully on the rack, and particular care must be taken not to damage the threaded ends and shoulders. Thread protectors should be used for both the box and pin ends when drill collars are stored on the rack and when they are moved across the pipe walk. When drill collars are ready to be picked up from the V-door, lifting subs or nipples must be made up tightly before the elevators are fastened. After each drill collar is picked up by the elevators, the pin thread protector can be removed.

Both pin and box threads must be carefully cleaned, all rust preventive removed, and a good coating of drill collar thread compound applied to the threads and shoulders. The last swipe of the dope brush should be on the box shoulder.

Mating threads of the pin and box should be stabbed carefully to prevent any damage to the threads or the box shoulder during lowering. Drill collar threads should be made up using hand tongs (fig. 4.23) rather than using the spinning chain. New threads should be made up by hand at least twice before tightening, to minimize thread galls on the initial makeup. Using both tongs, the threaded connection should be carefully tightened to the specified torque for the drill collar size and thread type.

After a single drill collar is made up in the rotary, the assembly can be picked up with the elevators and lowered into the hole. If the collars are smooth, a safety clamp should be installed before releasing the elevators. If the collars have slip recesses, the clamp is not needed. Grooved drill collars can save time while tripping, but OD wear can create a hazardous condition if the elevator shoulders become worn. Drill collars should never be broken out with the rotary. Tongs should always be used.

Figure 4.23. New collar being made up by hand with chain tongs

119

DRILL STEM AUXILIARIES

Various auxiliary tools are used with the drill stem, including drill stem subs, vibration dampeners, lifting subs, stabilizers, reamers, pipe wipers, and protectors. All should receive proper care and regular inspection.

Drill Stem Subs

A sub, or substitute, is a short threaded piece of pipe used to connect parts of the drilling assembly for various reasons (fig. 4.24). Most subs have box and pin threads, but double box subs can be obtained. Size and type of all subs should be identified by color coding or stencil markings.

The *kelly saver sub* is so called because its use minimizes wear of the threads on the kelly; it is often called the kelly sub. It is attached to the kelly, and into it is screwed the top length of drill pipe. A rubber casing protector is usually used on the kelly sub to reduce wear on the kelly and the top of the well casing.

A *crossover sub* is used between two sizes or types of threads in the drill stem assembly. A drill collar sub serves the same purpose but fits between the drill string and the assembly of drill collars. A *bit sub* serves as an adapter between the drill collars and the drilling bit.

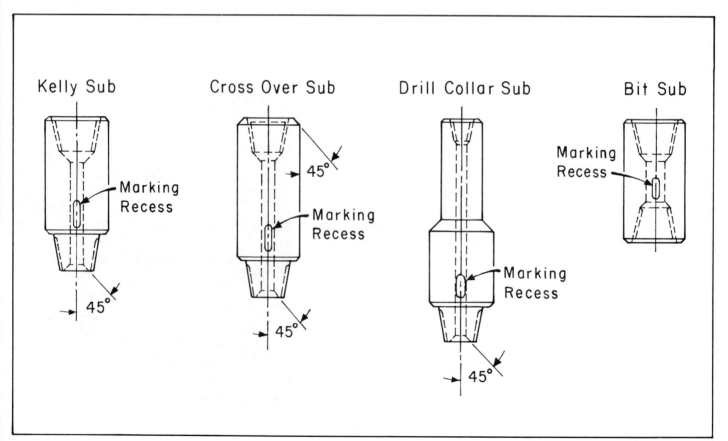

Figure 4.24. Drill stem subs

Vibration Dampeners

Vibration dampeners, sometimes called shock subs, are employed in the drill collar assembly, usually just above the bit, to compensate for the bounce and vibration of a roller cone bit as the cones rotate. Vibration dampeners may contain rubber, springs, compressed gas, or other springing elements for absorbing the bouncing motion of the bit. Vibration dampeners are designed to work like shock absorbers, permitting normal drilling without subjecting the drill stem and bit to damaging bounces.

Lifting Subs

Lifting subs (fig. 4.25) are not part of the drill stem but are usually made up into a stand of drill collars to provide a shoulder so that the drill collars can be picked up by the elevators and set back. Lifting subs may be provided with either square or 18-degree (bottleneck) shoulders to enable drill collars to be handled without elevators being changed. The threaded pins must match the drill collar box threads. Lifting subs must be made up securely when pulling out of the hole, using both sets of tongs.

A common problem with subs is that they may be made with substandard materials, or they may be poorly threaded. The steel may be soft and wear out quickly, causing serious damage to the tool joint boxes in the drill string. All sub threads should be protected with a clean thread protector. Subs on the rig should be properly racked and kept off the rig floor.

Figure 4.25. Lifting subs

Stabilizers and Reamers

Stabilizers are usually made up in the assembly of drill collars to stabilize the bit and the drill collars in the hole. Reamers are used to enlarge an undergauge hole.

Pipe Wipers and Mud Boxes

A pipe wiper is a rubber ring that fits around the pipe and cleans mud off as the pipe comes out of the hole. It also keeps junk from falling into the hole during tripping in. The mud box is a case that closes around a stand of wet pipe coming out of the hole and helps to keep mud from spewing over the crew and rig by directing the mud back into the pits.

Protectors

Thread protectors are used on drill pipe and drill collars to protect threads and shoulders from damage. Stabbing protectors guide the stand of drill pipe that is being stabbed to make a connection. They protect tool joint threads and shoulders. Pipe protectors keep tool joints from rubbing against the casing.

OPERATIONS INVOLVING THE DRILL STEM

Going in the Hole

The tool joint pin should be stabbed without striking the box shoulder. Before tool joints are spun up, alignment should be checked. If the joint wobbles and binds, high-speed rotation can burn threads. After spinning up, the rotary tongs should be used to tighten the joint to the recommended torque. Makeup should not be forced; improperly engaged threads can cause thread galling.

Tonging tool joints properly is the most important single factor in preventing tool joint difficulty. The proper torque should be achieved for the size, type, and class of tool joint used. Torque-measuring equipment should be used to guard against excessive makeup torque. Torsional failures in the rotary result from overtonging or from the use of slick thread compound, particularly with small tool joints.

It is good practice to use both tongs when making up or breaking out a tool joint. Washing occurs if the connection is not tightened to provide a proper shoulder seal and hold the hydraulic pressure. Leakage at this point will flush the lubricant from the threads and soon cause a washout. The kelly sub mates with every box in the drill string. Therefore, the threads of the kelly sub must be in good condition to avoid damaging every joint of drill pipe. Damaged tool joints should not be run in the hole.

Coming out of the Hole

When coming out of the hole, just as when running in, the slips should be set *after* the pipe has been stopped. Otherwise, large shock loads concentrated in the slip area can compress or stretch the pipe. Both tongs should be used to break out the tool joint. The slips should not be set too low, and backup tongs must always be used to prevent bending in or near the slip area. When the threads are disengaged, enough tension should be maintained on the spring hook to lift the tool joint pin from the box. The pin must not be allowed to strike the box when the spring hook jumps it out. The joints should be pushed to the side to prevent the pin from striking the shoulder when it drops down.

A good procedure is to come out of the hole on a different break each trip so that every connection can be broken periodically, its condition observed, and makeup torque checked. This may prevent a wobble and subsequent washout. The pipe setback area should be washed clean of mud as soon as each row of stands is placed.

When laying down the drill string, the following procedures should be followed.

(1) Wash the tool joints and the drill pipe inside and outside with clear, fresh water.

(2) Be sure the threads and shoulders are lubricated, preferably with a rust preventive if the drill string is to be stored for a long time.

(3) Install thread protectors, preferably before laying down the pipe.

(4) Check the drill string for straightness. Set aside any crooked joints so that they can be straightened. Crooked pipe causes wear on both the tube and tool joints.

(5) When laying down pipe, keep the tool joints from striking other pipe or metal of any kind, either on the ramp or on the pipe walk.

(6) When laying down, loading on trucks, and handling on the pipe rack, avoid leaving any length of pipe unsupported.

When making trips, particularly when pulling out of the hole, the crew should watch for the following conditions:

(1) *Dry or muddy threads.* Dry or muddy threads indicate that there is no shoulder seal. Even a slight shoulder leak causes fluid pressure inside the drill stem to force out thread lubricant and eventually lead to a washout.

(2) *Gall spots or bruised places on the shoulders.* Both gall spots and bruises on the shoulders eventually form a high spot, which will probably cause wobbling or leaking. Wobble on two opposite high places usually breaks the threads adjacent to the high places and wears those threads at 90 degrees to the high places.

(3) *Torsional damage and failures.* Excessive torque necks down threads by stretching the pin. The pin may part due to excessive torque. Excessive makeup torque often occurs downhole on connections improperly made up in the rotary or in the mousehole.

Critical Rotary Speeds

Critical rotating speeds in drill strings cause vibrations which may cause crooked drill pipe, excessive wear, and fatigue failure. Critical speed varies with the length and size of the drill stem and with the hole diameter. Excessive power is required for the rotary to maintain speed at critical levels. This power indication plus surface evidence of vibration warns the crew that rotary speed is in the critical range.

LESSON 4 QUESTIONS

Put the correct answer in the blank before each question. If there is more than one correct answer, put in all the correct letters. If a blank is drawn in the question, write out the answer as well as supply the letter in the multiple choice slot. The very act of writing down the answer will help you remember it.

Look again at Lesson 4 from the **Introduction** up to **Tool Joints.** Answer the following questions:

_____ 1. The drill stem includes –
      A. the swivel.
      B. the kelly.
      C. drill pipe.
      D. drill collars.
      E. all of the above.

_____ 2. The drill string includes –
      A. the swivel.
      B. the kelly.
      C. drill pipe.
      D. drill collars.
      E. the bit.

_____ 3. The American Petroleum Institute standardized threads and fittings so that –
      A. all wells could be drilled by a single company.
      B. equipment from different manufacturers could be used together.
      C. cheaper materials could be used.

_____ 4. Prior to 1930, drill pipe was run in compression to put weight on the bit.
      A. True
      B. False

_____ 5. Drill collars –
      A. reduce tool joint failures.
      B. lighten the drill string.
      C. improve straight-hole drilling capability.
      D. do all of the above.

_____ 6. Torsional strength can be most improved by changing to drill pipe with –
      A. greater inside diameter (ID) and smaller outside diameter (OD).
      B. greater OD and smaller ID.
      C. greater OD and ID.
      D. smaller OD and ID.

_____ 7. The amount of force required to cause permanent distortion of drill pipe is its

_____.
   A. minimum yield strength
   B. maximum torsional strength
   C. minimum tensile strength
   D. none of the above

_____ 8. The type of drill pipe having a constant inside diameter is known as

_____.
   A. internal upset
   B. internal-external upset
   C. a drill collar
   D. none of the above

_____ 9. Range 2 drill pipe is _____ feet long.
   A. 10 to 13
   B. 18 to 22
   C. 27 to 30
   D. 38 to 45

Questions 10–13, offered for extra credit, can be answered by using figure 4.3, concerning torsional and tensile yield strengths.

_____ 10. A linear pull of more than _____ lb on 4-foot pipe tongs would twist off 3½-inch, grade 95(X), 13.30 lb/ft drill pipe.
   A. 3,250
   B. 4,585
   C. 5,875
   D. 7,525

_____ 11. You could safely apply 70,000 ft-lb of torque to 5-inch drill pipe of grade

_____ with a nominal weight of _____.
   A. 95(X); 16.25 lb/ft
   B. 105(G); 25.60 lb/ft
   C. 135(S); 19.50 lb/ft

_____ 12. Using grade E, 4-inch, 14.00-lb/ft pipe and 35,000-lb drill collar weight, you could

safely drill to _____ feet with a 50,000-lb overpull safety factor.
   A. 14,000
   B. 18,000
   C. 22,000
   D. 24,000

_____ 13. The torsional strength of the tool joint is nearly always _____ than that of the drill pipe tube.
   A. more
   B. less

_____ 14. Factors affecting tool joint OD wear do not include –
  A. drilling weight.
  B. thread lubrication.
  C. rotary speed.
  D. abrasive formations.

_____ 15. One way to extend the drill string length safely with a given grade of pipe is to –
  A. use extra drill collars.
  B. overtighten joints.
  C. use longer drill pipe.
  D. use heavier pipe at the top of the drill string.

_____ 16. Collapse resistance may be an important consideration during a drill stem test (DST) because –
  A. drilling fluid is pumped at high pressure down the drill stem.
  B. during certain types of DSTs, the drill pipe above the testing tool is empty.
  C. lower-grade pipe is used during a DST.
  D. the pipe may come under tension if the packer is stuck.

_____ 17. During a DST of a high-pressure gas zone, _____

  _____ may be more important than collapse resistance.
  A. burst strength
  B. torsional strength
  C. nominal weight
  D. outside diameter

Questions 18–22 may be answered by using data from figure 4.4:

_____ 18. Grade E, 4½-inch, 13.75-lb/ft drill pipe might collapse if subjected to a

  hydrostatic pressure of _____ psi.
  A. 5,200
  B. 6,070
  C. 7,200
  D. 8,400
  E. 12,960

_____ 19. If you run a DST with empty 5-inch, 16.25-lb/ft pipe with bottomhole pressures

  reaching 9,100 psi, you should use new pipe of grade _____ or
  higher.
  A. E
  B. 95(X)
  C. 105(G)
  D. 135(S)

_____ 20. In order to avoid bursting new, grade 95(X), 4-inch, 14.00-lb/ft pipe, internal

pressure should not exceed _____ psi.
A. 9,960
B. 13,720
C. 14,380
D. 22,440

_____ 21. Grade E, 5½-inch, 19.20-lb/ft pipe may burst if internal pressure exceeds

_____ psi.
A. 4,810
B. 6,070
C. 7,250
D. 9,190

_____ 22. The most common type of drill pipe failure is _____

_____.
A. bursting
B. collapse
C. fatigue break
D. none of the above

_____ 23. A drill pipe break caused by corrosion is due to _____

_____.
A. pure fatigue
B. notch fatigue
C. torsional fatigue
D. none of the above

_____ 24. Metal is weaker under static conditions than under working loads.
A. True
B. False

_____ 25. If stress is kept under a certain limit, steel may withstand _____

_____ cycles of dynamic loading.
A. 10,000
B. 144,000
C. 10,000,000
D. 75,000,000
E. an infinite number of

_____ 26. Drill pipe fatigue limit is determined by –
A. surface finish.
B. chemical composition.
C. tensile strength.
D. none of the above.

_____ 27. Fatigue strength in drill pipe steel is considered to be approximately

_____ its tensile strength.
   A. one-quarter
   B. one-half
   C. twice
   D. five times

_____ 28. The most critical cyclical stresses on drill pipe are –
   A. tension.
   B. compression.
   C. torsion.
   D. bending.
   E. tension and torsion.

_____ 29. A stress riser is –
   A. a device used to distribute stress throughout the drill string.
   B. a mechanical or metallurgical defect which concentrates stress.
   C. an upper-level platform on the drilling rig.
   D. none of the above.

_____ 30. Fatigue failure due to cyclical stress is always due to crooked pipe.
   A. True
   B. False

_____ 31. A notch within 20 inches of a tool joint is critical because –
   A. maximum bending occurs in that area.
   B. a notch concentrates stress.
   C. torsional forces may break the threaded connection.
   D. of all of the above.

_____ 32. A longitudinal notch is more dangerous than a transverse or circumferential one.
   A. True
   B. False

_____ 33. Tongs should be applied to the tool joint instead of the body of drill pipe because –
   A. tong marks which deviate even slightly from the vertical tend to be stress risers.
   B. the drill pipe is more likely to bend if tongs are applied to the pipe body.
   C. the tongs are more likely to crush the pipe body.
   D. of all of the above.

_____ 34. Tong marks are the most common cause of notch fatigue failure.
   A. True
   B. False

_____ 35. Both sets of tongs should be used when making up or breaking out drill pipe –
A. because one set of tongs is more likely to bend the pipe.
B. so that in case one set fails, the other can be used as backup.
C. because the drill pipe may fall through the rotary table if only one set is used.
D. to avoid scarring of the pipe by the slips.

_____ 36. Insufficient makeup torque may cause –
A. galled tool joint shoulders.
B. last-engaged thread failures.
C. excessive downhole joint tightening.
D. all of the above.

_____ 37. Using the rotary table and slips to apply torque in making or breaking a connection can result in –
A. transverse scoring of pipe.
B. bending.
C. a loose connection.
D. an overtight connection.
E. all of the above.

_____ 38. If you use the slips to stop downward motion of the pipe, you may cause –
A. overtorquing of connections.
B. necked drill pipe.
C. excessive loads on the rotary table.
D. excessive loads on the elevators.
E. all of the above.

_____ 39. Corrosion fatigue may be caused by –
A. hydrogen sulfide.
B. carbon dioxide.
C. notch failures.
D. acid mud.
E. any of the above.

Now review the section on **Tool Joints** and answer the following questions.

_____ 40. Nearly all tool joints made today are –
A. removable.
B. shrink-on.
C. hardfaced.
D. flash-welded.
E. none of the above.

_____ 41. The type of drill pipe is identified on _____

    _____.
    A. the drill pipe body
    B. the pin taper
    C. the tool joint shoulder
    D. all of the above

_____ 42. Ordinary greases and oils should not be substituted for thread compounds –
    A. because they do not contain soft metallic fillers.
    B. unless the working temperature is below 40°F.
    C. because thread compounds help in applying proper makeup torque.

_____ 43. Using tubing or casing thread lubricants on tool joints may cause –
    A. overtightening.
    B. undertightening.
    C. swelled boxes.
    D. all of the above.

_____ 44. _Wobble_ –
    A. is movement between the surfaces of the mating box and pin.
    B. can be caused by improperly mated shoulders or low makeup torque.
    C. is used to drill at an angle.
    D. occurs only at the bottom of the drill string.

_____ 45. Dry or muddy pins can indicate _____.
    A. undertorquing
    B. wobble
    C. crowning
    D. mud pressure leakage
    E. all of the above

_____ 46. The pin shown in figure 4.14 should be –
    A. cleaned and lubricated before reuse.
    B. mated with an undamaged box in the drill string.
    C. lubricated with a special heavy-duty thread compound.
    D. none of the above.

The following questions cover material from **Drill Collars.**

_____ 47. Drill collars are not used to –
    A. strengthen the upper end of the drill string.
    B. apply weight to the drill bit.
    C. apply tension to the drill string.
    D. maintain vertical hole.

_____ 48. Bore or inside diameter of a drill collar is larger than that of drill pipe.
    A. True
    B. False

_____ 49. Which of the following sentences describes the relationship between drill collar weight and mud weight?
   A. Drilling in heavy mud requires fewer collars than drilling in light mud.
   B. Light mud provides more buoyancy than heavy mud, requiring the use of heavier drill collars.
   C. Drill collars weigh less in heavy mud than in light mud.
   D. None of the above.

_____ 50. Using 15 percent excess drill collar weight over the amount needed on the bit –
   A. is unsafe.
   B. is standard practice.
   C. causes excessive bit wear.
   D. compresses the drill string.

_____ 51. Drill collars help drill vertical hole because –
   A. they resist bending.
   B. they tend to hang vertically due to gravity.
   C. they apply tension to the drill string, preventing buckling.
   D. of all of the above.

_____ 52. The longer the section of drill collar string between the bit and the lowest point touching the side of the hole, the greater is the tendency to drill vertical hole.
   A. True
   B. False

_____ 53. A stiff drill collar assembly is less effective in drilling vertical hole than a limber one.
   A. True
   B. False

_____ 54. The bit can be stabilized by using –
   A. square drill colars.
   B. a packed-hole assembly.
   C. flexible drill collars.
   D. reamers.
   E. any of the above.

_____ 55. Why does most drill collar failure occur in the threaded connection?
   A. Drill collar tool joints are smaller in diameter than those in drill pipe.
   B. Drill collars flex more than drill pipe.
   C. The threaded connection is made of different materials than the collar body.
   D. Most bending occurs there because the collar body is stiffer.

_____ 56. In one rotation, each fiber in a bent drill collar goes through a tension-compression cycle.
   A. True
   B. False

_____ 57. Which of the following statements are true?
   A. Fatigue cracks originate where bending stress is highest.
   B. Fatigue cracks enlarge under tension.
   C. Corrosion fatigue may lead to notch failures.
   D. All of the above.

_____ 58. Signs of a loose joint may include –
   A. a dull gray color on the joint shoulder.
   B. galling.
   C. a dry pin.
   D. a swelled box.
   E. all of the above.

_____ 59. In drill pipe and drill collar connections, the threads form a continuous seal.
   A. True
   B. False

_____ 60. A loose joint usually fails –
   A. while tripping out.
   B. while performing a drill stem test.
   C. near the base of the pin.
   D. during a blowout.

_____ 61. A swelled or belled box may indicate –
   A. overtightening.
   B. joint size too large relative to drill collar size.
   C. incorrect thread lubricant.
   D. any of the above.

_____ 62. If a drill collar is worn to a smaller OD, makeup torque should be increased.
   A. True
   B. False

_____ 63. Drill collars with slip recesses –
   A. must be secured with safety clamps.
   B. can be broken out with the rotary.
   C. can be tripped out more quickly.
   D. should be made up using hand tongs.

Review **Drill Stem Auxiliaries** and **Operations Involving the Drill Stem** to answer the following questions.

_____ 64. A kelly saver sub –
   A. is used in place of the kelly.
   B. is made up between the kelly and the drill string.
   C. reduces wear on the kelly pin.
   D. reduces wear on the drill string.

_____ 65. A crossover sub is used to make the bit drill at an angle.
  A. True
  B. False

_____ 66. A device used to reduce damage caused by the bit bouncing off the bottom of the hole is called –
  A. a bit sub.
  B. a kelly sub.
  C. a vibration dampener.
  D. a shock sub.
  E. none of the above.

_____ 67. You can use standard elevators to handle drill collars while tripping out if you use

  _____.
  A. the kelly
  B. a kelly saver sub
  C. a lifting sub
  D. a crossover sub
  E. none of the above

_____ 68. A reamer may be used –
  A. in a packed-hole assembly.
  B. to enlarge undergauge hole.
  C. with a stabilizer.
  D. in all of the above cases.

_____ 69. The mud box is where mud is stored while tripping out.
  A. True
  B. False

_____ 70. When coming out of the hole –
  A. the pin joint being broken out should be pushed aside to avoid bouncing against the box shoulder.
  B. a different connection should be broken each trip to observe joint condition and distribute wear among all the joints.
  C. both tongs should be used to break out each tool joint.
  D. the tool joint should be set low to avoid bending.
  E. all of the above are true.

_____ 71. Excessive makeup torque may occur downhole on a loose connection.
  A. True
  B. False

_____ 72. Signs of critical rotary speed may include –
  A. vibration.
  B. excessive penetration rate.
  C. excessive power requirement.
  D. bent drill pipe.
  E. all of the above.

Lesson 5
DRILLING BITS

Introduction

Roller Cone Bits

Diamond Bits

Drag Bits

Lesson 5
DRILLING BITS

INTRODUCTION

The bit and its performance are what rotary drilling is all about. When the bit is on bottom and making hole, it is making money – but only as long as it is an effective cutting tool. To be an effective cutting tool, the bit must be in good condition. Weight must be applied to make the bit drill, and supplying this weight is one function of the drill collars. The bit must be rotated, and rotation is the function of the drill stem and the rotary. Finally, the drilling fluid must cool and lubricate the bit as it removes chips and cuttings from the bottom of the hole.

Many variables affect bit performance, particularly the type of formations being drilled. These variables usually involve questions of economics, especially in the selection of a bit type that can drill most economically. Drillers want a bit that has a good rate of penetration (ROP), lasts a reasonable number of rotating hours, and drills holes the same size as the bit (true-to-gauge). Essentially, the driller is looking for a bit that averages the most feet per hour and lasts the most hours possible. If the sides of the bit wear down, it will drill an undersize, or undergauge, hole. Out-of-gauge holes not only cause lost time for reaming, but can also stick the drill stem, cause a fishing job, and thus increase drilling costs.

In most situations, the objective is to get all the footage possible from a bit, thereby minimizing the number of round trips needed for bit changes. Situations occur in which only one or two bits are needed before pulling out for a survey running casing. For example, when making hole to set surface casing in extremely soft formations, only one bit may be needed, and occasionally, that bit may be used for several wells. Deeper drilling in harder rocks is more difficult. Trip time increases, and the driller will generally run a bit designed to drill these formations as well as to drill the surface casing cement left in the hole. In some instances, the cement may be drilled and then bits changed to drill that particular formation.

Formations vary in hardness and abrasiveness. If the bedded strata did not change, one bit best suited for that formation could be selected, and drilling ahead could begin. Usually, however, the strata are made up of alternating layers

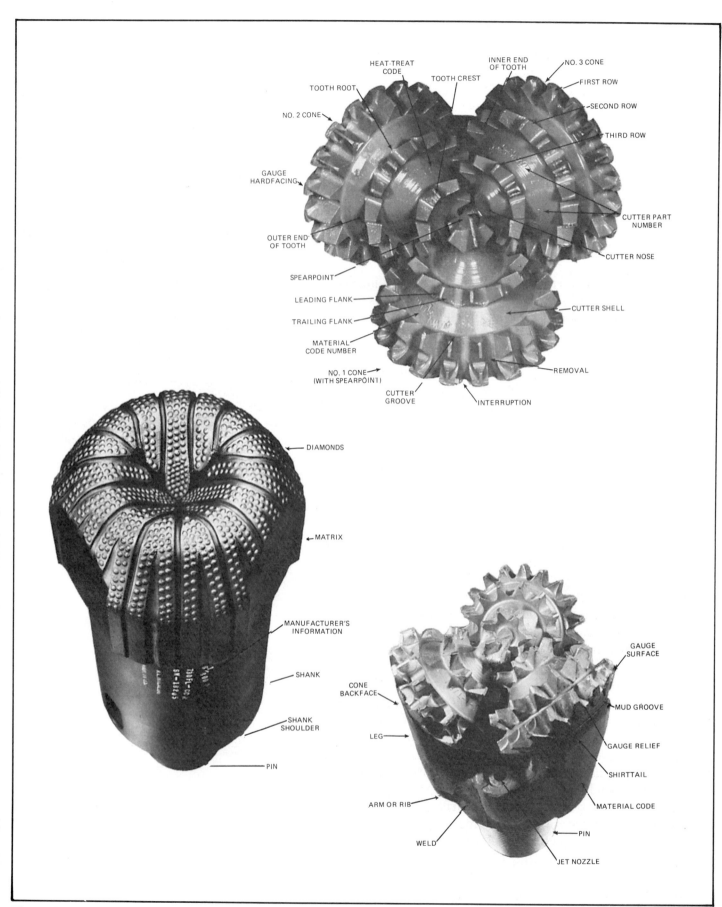

Figure 5.1. Bit terminology

of soft material, hard brittle rocks, and hard, abrasive sections. Instead of changing bits every time a new type of formation is encountered, a compromise bit that can drill through all types of formation is chosen. If drilling is taking place in a known field, information about the formations is available to the driller. If a wildcat well is being drilled, a process of trial and error will have to be followed in selecting bits.

Three general types of bits will be discussed in this lesson: roller cone, diamond, and drag bits. *Roller cone*, or *rock*, *bits* have cone-shaped steel devices called cones that turn as the bit rotates (fig. 5.1). Most roller cone bits have three cones, although some have two and some have four. Teeth can be cut out of the cones, or very hard tungsten carbide buttons can be inserted into the cones. The teeth or tungsten carbide inserts will actually cut or gouge out the formation as the bit is rotated. *Diamond bits* do not have cones or teeth (fig. 5.1). Instead, several diamonds are embedded in the bottom and sides of the bit. Diamonds are so hard that diamond bits can drill quite efficiently. *Drag bits* are used to drill shallow, soft formations close to the surface. All bits have passages drilled through them to permit drilling fluid to exit.

ROLLER CONE BITS

The first roller cone bit was introduced to the industry in 1909. For the next fifteen years it gradually but steadily gained acceptance, being used primarily in hard-formation areas.

The two cones on early rock bits did not mesh; the teeth of one cone did not fit into the spaces between the teeth of the other. So they readily balled up in soft shale. When they needed attention, roller bits were dressed on the rig floor with new cones and reaming rollers. Also, the drilled watercourses – passageways through the bit for the drilling fluid to get out – washed out quickly. A replaceable wash pipe, which could be renewed when the bit was dressed, solved that problem.

Rock bits were constantly studied and improved. The Hughes Simplex rock bit was redesigned with meshing teeth for self-cleaning. It had a reservoir for lubrication (fig. 5.2). During the 1930s, the three-cone rock bit was introduced. It featured bearings lubricated by drilling mud and cutters designed for various formations. These bits looked similar to the rock bits available today, but they did not have certain improvements. The following four developments should be noted:

(1) the change of watercourses to the jet bit arrangement;
(2) the introduction of tungsten carbide inserts to replace steel teeth;
(3) the use of lubricated sealed bearings; and
(4) the use of journal (friction) bearings.

Figure 5.2. Hughes Simplex rock bit of mid-1920s

Figure 5.3.
Soft-formation
milled-tooth bit

Figure 5.4.
Hard-formation
milled-tooth bit

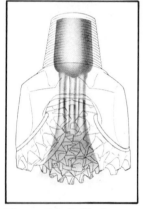

Figure 5.5. Drilled
watercourses in a
regular bit

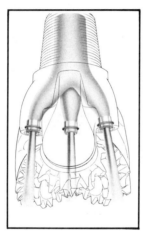

Figure 5.6. Fluid
passageways in a
jet bit

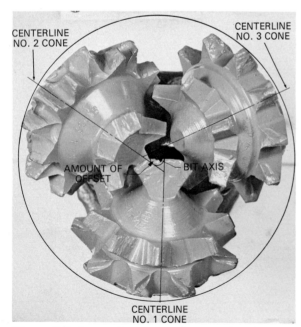

Figure 5.7. Cone offset in a soft-formation bit

Design

Bit designers have produced a bit with long, widely spaced teeth on a thin shell with small bearings to be used on soft formations (fig. 5.3). To overcome balling up of soft-formation bits, they have developed interrupted or deleted gauge-row patterns, so that cuttings, being smaller than the space between teeth, do not get stuck. For medium formations, they have designed teeth somewhat shorter than those for soft formations, yet fairly well spaced. Hard formations require bit teeth shortened even more for added strength; they chip, rather than scrape, the formation (fig. 5.4).

Designers are also concerned with fluid passageways in the bit. Regular or conventional watercourses (fig. 5.5) were the first to be integrated into a roller cone bit and are still in limited use today. These courses direct drilling fluid onto the cutters to keep them clean, and the fluid then goes on to clean the hole. The speed or velocity of the stream of drilling fluid in this type of passageway is relatively low, and the disadvantages of the system include balling and cone erosion. Such conventional watercourses have been almost completely replaced by jet watercourses, or nozzles (fig. 5.6). These nozzles direct the stream of drilling fluid past the cones and completely flush out cuttings in the hole. The stream of drilling fluid can be controlled by changing the nozzle size and can improve the rate of penetration in soft formations by washing away or eroding the formation even before the bit touches the bottom of the hole. One company now offers a bit that has two cones and extended nozzles.

Rock bits may be classified in general as (1) steel-tooth (milled tooth) bits and (2) tungsten carbide insert bits.

Steel-tooth (Milled) Bits

Early rock bits were made with two cones that simply rolled wheelfashion on the bottom of the hole as the bit rotated. These cones were perfectly aligned with each other, in contrast to cones of modern bits. This design is still followed on most hard-formation types of bits; in drilling hard formations, the action of the teeth is simply to crush and, to some extent, chip the rock formation in order to make hole.

Modern soft-formation rock bits have their cones offset (fig. 5.7). The off-center alignment of the cone makes the teeth scrape and gouge the formation as the cone rolls on the bottom of the hole, the amount of scraping depending on the amount of cone offset. Intermeshing of the teeth makes it possible to use longer teeth as well as to include a self-cleaning effect. Bits designed for the softest formations with the least amount of abrasive characteristics have the most cone offset. Reduced offset (or none) is used in bits intended for the hardest, most abrasive formations. In general, widely spaced, long, sharp teeth are used for soft formations, and more closely spaced, shorter, stronger teeth are used for harder rocks. Steel-tooth bits are available for soft, medium, and hard formations, although tungsten carbide insert bits are rapidly replacing hard-formation milled bits. Bits that have self-sharpening teeth are also available; on these, one side of the tooth is hard-faced and the other is not, so when the softer side wears away, the hard side still holds a point.

Tungsten Carbide Insert Bits

In recent years, many improvements have been made in sealed-bearing tungsten carbide insert bits. In the past, these bits had only small carbide extensions, adequate for slow penetration rates being achieved at the time. Carbide insert bits were used to reduce trip time, because the same bit could be used on different formations; however, slow bit speeds reduced the rate of penetration and faster speeds could cause insert breakage.

Modern tungsten carbide insert bits run at high rotary speeds—up to 180 revolutions per minute (rpm) or more, as compared to 45 rpm used with older ones. The new high-speed carbide insert bits also run with much higher weight on bit (WOB) than the older ones, reducing the expected bit life in terms of hours run.

In the past, most carbide insert bit failures arose from broken cutting structures, cone-shell erosion, and bearing failure. In modern carbide insert bits, the jets afford more safety for inserts, sweeping away the cuttings faster and more thoroughly than conventional drilling fluid passageways. New developments in cone materials have made the cones more wear-resistant, cutting down on cone failure. The really big innovation in carbide insert bits has been the development of sealed bearings, since bearing failure was one of the more common failures of this type of bit.

The advantages of the carbide insert bit include great durability, good insert burial into the formation—up to 80 percent of the insert per revolution in soft formations—and the ability to drill different types of formations with the same bit. Their disadvantages include the fact that erosion around the base of inserts can result in their loss, and the possibility that with complete insert burial an area of the cone shell can come into contact with the formation and transmit shock loads from the drill string directly to the bearing.

The carbide insert bit has received such wide acceptance that research and development programs for it have been given priority, so it will probably continue to improve rapidly.

Many general types of steel-tooth and insert bits are now available (fig. 5.8). The comparative number and length of cutters on bits for soft, medium, and hard formations should be carefully noted. Insert bits can be obtained for soft to medium-hard formations having high compression strength and for hard, abrasive formations.

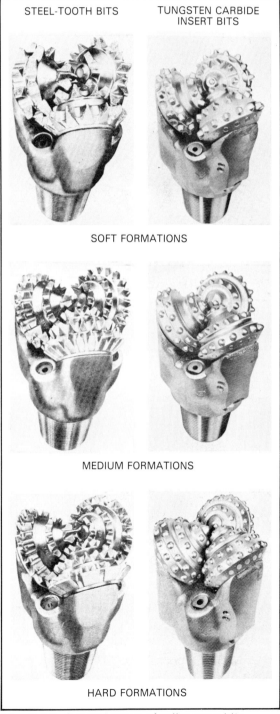

STEEL-TOOTH BITS   TUNGSTEN CARBIDE INSERT BITS

SOFT FORMATIONS

MEDIUM FORMATIONS

HARD FORMATIONS

Figure 5.8. General types of roller cone bits

Bearings

Rock bit bearings must give satisfactory service under rather severe conditions. For example, an 8¾-inch bit may be rotated at 35 to 70 revolutions per minute with 70,000 pounds of weight or rotated in excess of 250 rpm with 40,000 pounds of weight. Roller cone bits may be rotated for 30 hours or more in abrasive mud and up to 300 hours under optimum conditions.

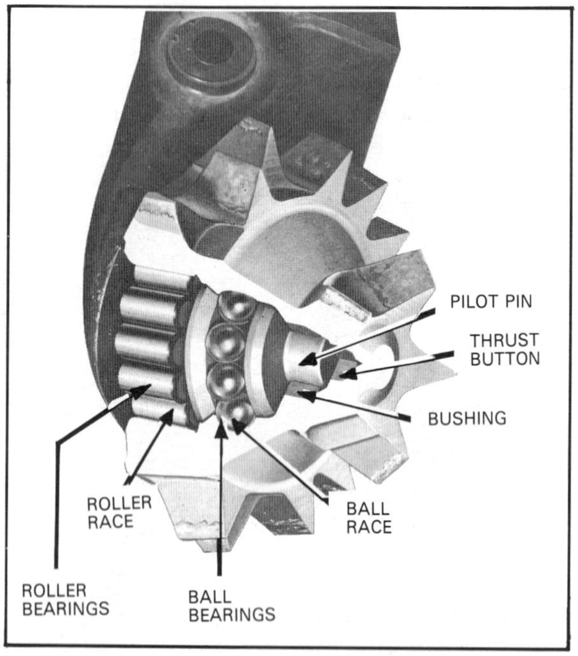

Lubrication. Lubrication is obviously of extreme importance to bits and bearings. In bits that do not have sealed bearings, drilling fluid provides bearing lubrication. However, sealed bearings are, as the name implies, sealed off so that drilling mud cannot reach them. Therefore, they must be provided with lubrication from another source. Usually, a lubricant reservoir is placed on each leg of the bit when it is manufactured.

Most of the lubricants used in bearings contain petroleum bases and graphite. Some contain other additives, even soap. Some bearings are self-lubricating; one such bearing relies on heat to lubricate itself. Silver, lead, and indium are used on surfaces that will come into contact with other surfaces. As the surfaces heat up from friction, the indium melts in small quantities that provide lubrication.

Roller bearings. A series of compromises must be made in designing roller bearings (fig. 5.9) until the various elements are properly balanced with each other so as to obtain the best overall performance. The number and size of rollers selected must provide a bearing pin roller race of sufficient size to prevent early fatigue failure. The maximum number of roller bearings should be used to reduce the unit loading so that spalling (flaking) and wear are postponed as long as possible. In addition, the diameter of each roller must be large enough to prevent breakage.

Sealed bearings. For many years rock bits were produced with nonsealed bearings only. This type of bearing had a relatively short life because abrasive drilling fluid could enter it and cause spalling and abrasive wear of all bearing elements, especially the bearing pin roller race. The first functional sealed-bearing rock bit was introduced to oil fields in 1959. This new bit made extended bearing life possible by permitting the bearing to operate in a clean lubricant environment.

The essential components of a sealed bearing (fig. 5.10) include bearing, seal, reservoir, and pressure compensator. The seal keeps the lubricant inside and the mud outside the bearing, while the reservoir provides the lubricant that is fed into the bearing to replace what is used. The pressure compensator maintains equal pressure inside and outside the bearing. The potential advantage of the sealed bearing is realized only if all parts of the system function properly.

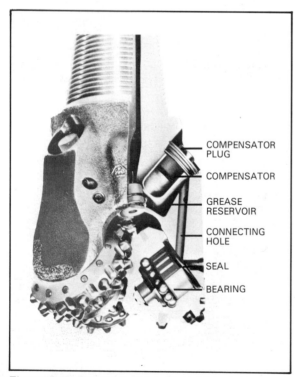

Figure 5.9. Regular, or nonsealed, bearings

Figure 5.10. Sealed and lubricated bearings

142

Sealed roller bearings, introduced first in carbide insert bits and later in steel-tooth bits, increased bearing life as much as one-fourth. Because the rollers developed spalling failures, journal bearings, in which line contact of rollers was replaced by area contact of journals, were developed. Journal bearings tend to gall rather than spall; therefore research became focused on developing features that would minimize friction.

All of the major manufacturers of rock bits now offer sealed, lubricated roller and journal bearings for maximum service life.

Journal bearings. A journal bearing tungsten carbide insert bit (fig. 5.11) has its cones retained on the bearing pin by balls, as roller bearing bits do. A ring bearing seal is used to protect against intrusion of drilling fluid and loss of the specially compounded grease. Replacing the rollers with an integral journal makes possible maximum pin strength through use of a larger pin diameter. A wear-resistant hard surface is used on the pressure side of the journal pin, and a special alloy or inlays of a special alloy, functioning as a solid lubricant, are used in the cone journal race for maximum resistance to galling. A specially compounded grease is used in the steel-tooth bits to resist galling.

Long bearing life has resulted in fatigue limits in areas not previously considered critical and has created a need for tungsten carbide inserts that are both more resistant to wear and less susceptible to breakage than previous inserts. The numerous stress reversals of long hours of rotation impose problems of insert retention. Possible improvements can be made in the fields of cone metallurgy, insert contour, press-fit tolerances and techniques, and surface finish of both the tungsten carbide inserts and the sockets. Journal bit bearings now have nearly unlimited life under normal use if all the support systems continue to operate.

Drilling Fluid Passageways

Nearly all modern bits are provided with jets instead of the conventional watercourses because jet nozzles are replaceable and can be changed to match the pressure and volume requirements of the drilling fluid. Jet bit nozzles are made of special erosion-resistant material. Fluid velocity through jet bit nozzles usually exceeds 250 ft/sec, depending on the diameter of the nozzle. Nozzle sizes are described in thirty-seconds of an inch. For example, a 10 nozzle is 10/32 inches in diameter. Figure 5.12 is an illustration of the standard and shrouded types of nozzles offered by one manufacturer. Spring-snap rings are used to retain the jet nozzles in the bit body.

Weight and Rotary Speed Factors

Experience with steel-tooth bits has proven that drilling rates in brittle rocks increase more than proportionately to the increase in drilling weights. A 30 to 40 percent increase in

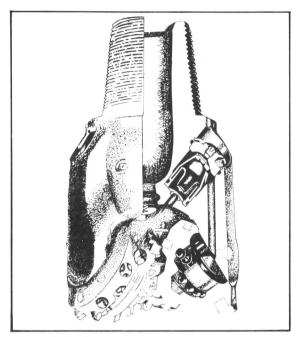

Figure 5.11. Journal bearings

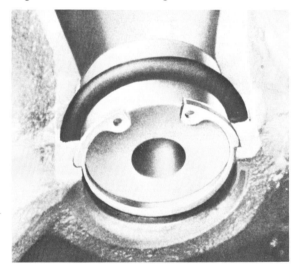

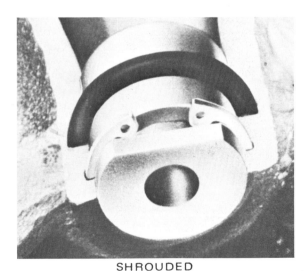

SHROUDED

Figure 5.12. Replaceable nozzles for jet bits

weight on a steel-tooth bit sometimes doubles the rate of penetration. An operator is not interested in improving just the rate of penetration, but in the overall performance, so as to achieve the minimum cost per foot of hole drilled. Achieving maximum rock-bit performance depends on both tooth form and structure as well as on bearing life.

In soft formations, the use of heavy weights is restricted because of the tendency of the steel-tooth bit to ball up. It is possible to increase rotary speed to offset the reduced weight. Good cleaning by high fluid velocity allows fast penetration rates in these formations. How much the rotary speed can be increased is limited by the abrasive properties of the formation. For harder formations, heavier weights are needed.

Field testing has shown that moderate weights and high rotary speeds are best for steel-tooth bits in nonabrasive formations. Heavier weights and lower rotary speed, however, are better for slowly drilled, abrasive formations. The best combination of weight and rotary speed varies with different formations. Testing has also indicated that, when the cutting structure of the bit becomes dull, the weight should be increased. Changes in formation characteristics will affect the weight and rotary speed required for economical drilling.

Tungsten carbide insert bits deserve the same attention to weight and rotary speed, the principal difference being that under normal conditions the teeth of a tungsten carbide insert bit wear only slightly. Therefore, weight and rotary speed should be relatively constant as long as a formation remains constant.

Jet Nozzle Factors

The watercourse in a conventional bit is positioned to direct the drilling fluid onto the cutters to keep the cutters clean with the main force of fluid. In this type of bit, only part of the fluid power reaches the bottom of the hole to lift the cuttings up the annulus. Watercourse openings and the usual circulation rate are such that fluid velocity is approximately 100 feet per second (ft/sec). Jet bit nozzles, on the other hand, are positioned so that the jet streams hit directly on the bottom of the hole instead of on the cutters. With this arrangement, fluid turbulence is created around the bit to keep the cutters properly clean. Fluid velocities in jet bits often exceed 400 ft/sec.

A popular method of operating jet bits is to select sizes for nozzles, pump pistons, and liners that allow maximum horsepower (in terms of pressure and fluid volume at the bit) within the following limitations:

(1) maximum horsepower available for driving the pumps;
(2) maximum pump discharge pressure depending on the pressure rating of the pump;
(3) minimum annular return velocity (120 ft/sec is often sufficient); and
(4) minimum practical nozzle size (small sizes plug more easily than larger sizes).

Note

Abrasive properties of formations determine rotary speed

144

Experience has shown that a direct relationship exists between jet velocity and rate of penetration – the higher the hydraulic power, the faster the drilling. Weight on bit and rotary speed also must be maintained at optimum values. The amount of pressure that can be usefully employed for rotary drilling is limited. Pump maintenance costs may be uneconomical at excessive pressures. Drill pipe, swivel packing, and other rig items may wash out at elevated pressures.

Special-Purpose Rock Bits

Jet air bits (fig. 5.13) are designed to use air, gas, or mist as drilling fluid. Passageways allow circulation of a portion of the drilling fluid past the nonsealed bearings to cool and clean them. Screens over the air inlet ports prevent the bearings from becoming clogged with cuttings or other foreign material.

Two-cone bits (fig. 5.14) are used for core drilling or for other small-diameter holes.

Jet deflection bits (fig. 5.15) are sometimes employed for directional drilling in soft formations for sidetrack operations. The drill pipe and deflection bit are lowered into the hole, and the large jet is oriented to wash out the side of the hole in a specific direction.

Wear and Problems

Bearing failure or wear is a frequent reason for ending a bit run. *Outer bearing failure* can be caused by excessive rotation time, coupled with heavy weight. Signs of outer bearing failure are skid marks on the cones and one or two cones locked while the rest roll freely. Other causes of this type of bearing failure are abrasive materials in the drilling fluid, sulfur water in the drilling fluid, or hydrogen sulfide gas. Reduction of abrasive materials from the drilling fluid, decrease in number of rotating hours, and reduction of weight on bit may reduce this problem.

Inner bearing failure can occur when a bit is used for a formation that is too hard for it. The teeth wear down to a flat top and, in order to obtain a good penetration rate, more weight is added to the bit, causing inner bearing failure. Other causes may be excessive gauge rounding, or reaming an undergauge hole that is too tight. This problem may be partially or completely eliminated by using a bit designed for harder formations, reducing rotating time, decreasing rotary speed, or choosing a bit whose shape is better suited to the size of the hole being reamed.

A combination failure of both inner and outer bearings can occur from excessive weight, excessive rotary speed, reaming, excessive rotating time, or overly abrasive drilling fluid.

Bit teeth wear away sooner or later. However, evaluation of the wear patterns can indicate characteristics of unusual wear and suggest ways to eliminate or reduce it.

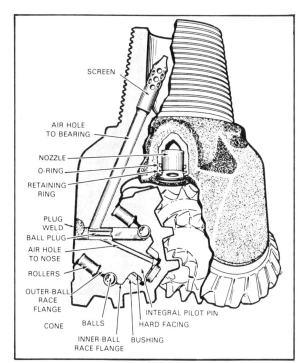

Figure 5.13. Cross section of a jet air bit

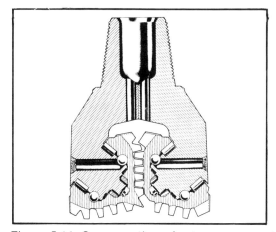

Figure 5.14. Cross section of a two-cone regular roller cone bit

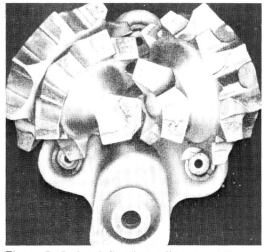

Figure 5.15. Jet deflection roller cone bit

145

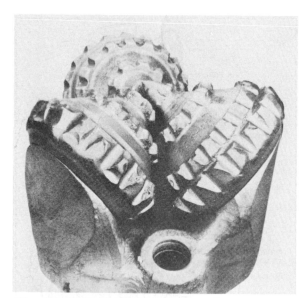
Figure 5.16. Excessive gauge rounding

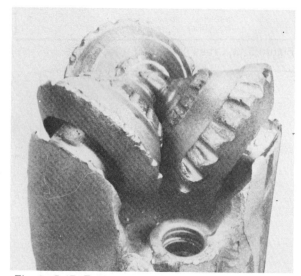
Figure 5.17. Extreme gauge wear

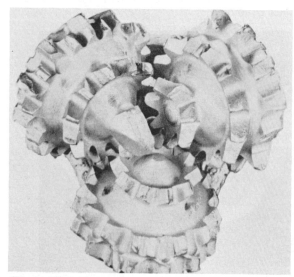
Figure 5.18. Excessive tooth breakage

One of the more serious problems encountered with bit wear is the *formation of an undergauge bit*. *Undergauge* means that the gauge teeth, or outermost teeth on the cones, have been worn away or chipped to such an extent that they are drilling a hole with a smaller diameter than is called for (fig. 5.16). This situation is serious because the next bit will have to ream the hole to the appropriate size, and such action may be harmful, as well as a waste of time and effort. In some extreme cases, the shirttail and gauge of the bit are so worn away that the rollers eventually fall out (fig. 5.17). This condition can be caused by using a bit that is not strong enough for the type of formation being drilled, running a bit too long, or using a rotary speed that is too fast for the specific bit.

Tooth chippage is fairly commonplace. It happens when the bit strikes a new type of formation that makes it bounce, chipping or breaking the teeth (fig. 5.18). Some tooth chippage is desirable in hard-formation bits and can be an indication that the right bit was chosen for the job.

Tooth abrasion is another type of wear occurring in regular drilling. In some bits, part of the tooth is hardened with special alloys and the other side is left without any special treatment; a self-sharpening effect is thereby produced. If tooth abrasion is excessive, the drilling fluid may be too abrasive, the circulation velocity may be too high, or the job may be more suitable for a jet bit. If a jet bit is being used and there is still too much abrasion, velocity or abrasiveness of the drilling fluid may be too high.

Bradding is a condition in which the weight on a tooth has been so great that the tooth has dulled until the softer inner portion of the tooth caves over the harder case area. This condition usually develops early in the run because of too much weight on bit or excessive bit speed and affects the inner rows or weaker teeth. Bradding can be associated with loss of teeth.

Tracking is a rare type of tooth wear. It occurs when the pattern made by all three cones on the bottom of the hole matches the bit-tooth pattern to such an extent that the bit gears itself to the formation and drills very little. Tracking usually happens in hard or medium formations and can be avoided by keeping proper weight on bit and maintaining proper rotary speed.

Off-center wear results when the bit gyrates rather than rotates around the true center of the bit, and allows ridges to develop in the hole. These ridges increase in size and wear away fronts and backs of teeth as well as cone shells. The condition is caused by improper bit use, weight, or speed. Using a bit for softer formation may remedy the problem.

Unbalanced tooth wear designates the breaking of inner, less massive teeth during the bit break-in period. It happens when the formation is too hard for the bit chosen or when too much weight or excessive bit speed is used. Using a bit for harder formations or decreasing bit weight or speed may alleviate the problem.

Balling up occurs when teeth become gluey with material from soft, sticky formations. This condition results from excessive weight on bit, which makes the teeth go down too far into the soft formation, or from a drilling fluid stream not strong enough to carry away the cuttings. Teeth at the bottom drag and become unduly worn. The remedy is to decrease weight on bit or to increase volume and/or velocity of drilling fluid.

Cone skidding occurs when a cone locks so that it will not turn when the bit is rotating (fig. 5.19). It results in a flattening of the surface of the cone in contact with the bottom of the hole.

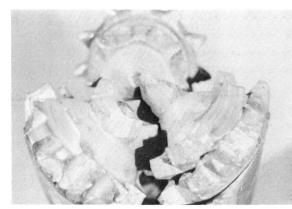

Figure 5.19. Bit with effects of cone skidding

Tungsten carbide insert bits show little wear on the cutting structures after normal operations. Most damage to this type of bit results from improper drilling practices or from abnormal conditions in the hole or formation. *Broken inserts,* the most common damage to insert bits, usually result from use of a bit that is not adequate for a hard formation. Insert bits for softer formations have longer inserts and a predesigned amount of cone sliding. Excessive loads on the inner teeth, produced when harder formations are encountered, readily cause broken inserts. Such a situation is best prevented by selecting a bit adjusted to the hardness of the formation expected to be encountered. But since various types of formations may be encountered, and tungsten carbide insert bits have such long runs, it may be necessary to use a softer-formation bit, increase the weight on bit, and decrease rotary speed when drilling through the harder formations.

Insert breakage can result from the impact of excessive bit speed used in harder formations. Breakage occurs in the area of greatest loading and impact force – the outside, or drive rows, of cones. If the bit undergoing breakage is a soft-formation bit, and a long run of hard formation is expected, the bit should be changed. If a hard-formation bit is being used, the bit speed should be reduced.

Insert loss by cone erosion (fig. 5.20) happens when the drilling fluid wears away the cone around the insert. The condition is usually produced by air or gas drilling, where excessive flow rates of the drilling fluid combine with abrasive cuttings to erode the cone. Very little can be done about this common problem in air or gas drilling.

One of the most frustrating problems faced in drilling with tungsten carbide insert bits is that, once the inserts start falling out or breaking, more damage is bound to occur from the broken pieces on the bottom of the hole. Tungsten carbide is so dense that it is almost impossible to circulate the broken pieces out of the hole; therefore, the pieces at the bottom damage cones or remaining inserts. Furthermore, it is very difficult to drill past past this junk in hard formations. Therefore, before a new bit is run into the hole after a damaged one is taken out, a junk basket should be run in and the inserts worked into the basket.

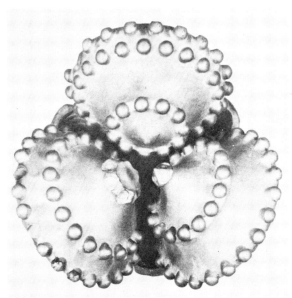

Figure 5.20. Severe cone-shell erosion of an insert bit

Figure 5.21. Cone erosion of an insert bit

Overgauge drilling

Cone problems usually result from poor drilling practices or improper cone design, and can be serious. *Cone erosion* (fig. 5.21) can be caused by excessively high circulation rates or abrasive drilling fluid. Severe erosion can cause a cracked cone, the crack most often occurring around the cone's circumference. Cracked cones can also result from fatigue cracking caused by tool marks, migrating cracks from stresses produced by inserts pressing into the cone, or shoddy workmanship in metallurgy or cone design.

Off-center wear often occurs when the penetration rate is too slow. The hole is not drilled fast enough to hold the bit, and the cone offset causes the bit to drill an overgauge hole, with two cones drilling the bottom of the hole and one drilling the side walls. Ridges form on the bottom of the hole and rub against the cones. Remedies for this situation include using stiff drilling assemblies to stabilize the bit, increasing penetration rates through heavier weight on bit, and selecting the proper bit. There will be times, however, when drilling or equipment requirements make it impossible to avoid off-center wear.

Center coring is a wearing away of the nose area of a cone. Excessive loads applied to this part of the cone cause such wear, and will eventually result in loss of inserts or cone breakage. There is really no remedy for this problem; however, it is extremely important to run the next bit carefully to clear the bottom of the hole of that portion of formation left by the missing cone-nose teeth.

Bearing failure can result in *locked cones,* which in turn cause broken inserts or teeth. The cones may shake, knocking the teeth of one cone into the matrix material of another and causing grooving of the second cone. If the action is allowed to continue, it may sever the nose of the cone completely.

The bit in figure 5.22 is considered to be in a desirable dull condition for a soft-formation bit, with a considerable amount of life remaining. There is little tooth breakage or chipping. The hardfacing has been effective in causing teeth to wear in such a manner that they remain relatively sharp, even though appreciably worn down in height.

It is important to note that most bit damage or wear is caused by improper weight on bit, high-speed drilling fluids, or bit selection, or by failing to take drilling conditions into sufficient account.

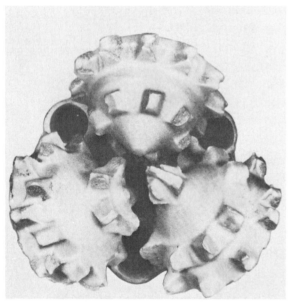

Figure 5.22. Desirable dull condition

Grading Bit Wear

Drilling bits must be properly graded for wear. Grading may affect the economics of drilling, the corrections of improper drilling practices, and the selection of proper bit type.

Teeth are graded in 1/8-inch intervals of the original tooth height. Using the letter T for teeth, *T3* means that 3/8 of the original height is worn away. For accuracy, a dull bit should be compared to a new bit. This practice is especially helpful when a change in bit type has been made.

The grading of a used bearing is the most difficult part of grading dull bits. Every piece of information about the bit should be used. Bearing life is expressed in ⅛-inch increments of bearing life spent. Using the letter *B* for bearings, *B8* means that the bearing is completely worn out, and *B6* means that ⁶⁄₈ of the estimated life has been used. The number of hours that a bit has been run should be known, as well as the operating condition under which it was used. For example, if a bit has been run 25 hours, and if the bearings are in such condition that another 15 hours can be obtained, then the bearings should be graded *B5* (that is, ²⁵⁄₄₀ or ⁵⁄₈ of its life has been spent). Bearings should not be graded unless it is known how long a bit has been run.

Figure 5.23 is a reproduction of a page taken from the *IADC Drilling Manual*, showing the symbols and methods used in the grading of dull rock bits.

Records and Costs

Rock bit manufacturers have collected records of bit performance covering large numbers of all types of bits run on all types of formations in all areas of the United States and many other countries. The manufacturers have relied on drillers' logs and bit records for this information. One type of bit record, kept on a West Texas well, is shown in figure 5.24. Note the preponderance of J-type (journal-bearing) bits. Bit records give details as to the size, make, and type of bit, size of jets, depth out, number of feet drilled, hours run, drilling weight, rotary speed, pump pressure and strokes per minute, drilling fluid weight and viscosity, and dull conditions of the bit when it was pulled. This information makes it possible for drilling contractors, toolpushers, and engineers to compile reports that can be used for future reference.

Bit runs can be compared by calculating the average cost of each foot of hole drilled, taking into account rig operating costs, the cost of the bit, and the round-trip time needed for a bit change. The following formula is employed:

$$C = \frac{B + OC\,(T + R)}{F}$$

where –

C = cost per foot of hole (dollars);
B = bit cost (dollars);
OC = rig operating cost per hour (dollars);
T = trip time (hours);
R = rotating time (hours of on-bottom drilling);
F = feet drilled by the bit.

Assume that the following information holds true for the bit record in figure 5.24:
Rig operating costs (OC) = $200 per hour.
Bit cost (B) = $900 for a steel-tooth, sealed bearing bit and $3,500 for an insert journal bearing bit.
Round-trip time (T) = 10 hours.

Systems of grading used roller bits have been adopted by the IADC, in 1963 on milled tooth bits and in 1971 on insert bits. This was to establish a uniform and rapid method by which the driller or toolpusher could report tne condition of a used bit. Bit records are very important in estimating the drilling costs of succeeding wells in a given area. They indicate the type bit to be used in a given formation to obtain the greatest footage per bit or per day and hour of rig time. The bit record should also include a statement of the condition of the bit when it is pulled. This makes it possible to select the proper bit for maximum footage.

To assist rig crews in reporting bit condition accurately and

quickly, some bit manufacturers are making available inexpensive mechanical aids to measure tooth and gage wear. Used in accordance with instructions from the bit manufacturers, the measuring devices greatly increase the accuracy of the bit wear reports. Since tooth wear is not uniform on any row of a given cone, it is advisable to take several readings and report an average figure. Where mechanical aids are not available, it is necessary to "eyeball" the tooth wear and estimate wear in "eighths" to be consistent with the system.

The "Bit Condition" of each bit should be reported in the "Remarks" section of the IADC-API Daily Drilling Report form and on the Bit Log or Record. Terminology is shown in Table A2-1.

CODE ABBREVIATIONS

GENERAL NOMENCLATURE

A	–	Apex
B	–	Broken or Bearing
C	–	Cone or Chipped
CR	–	Cored
E	–	Eroded
G	–	Gage
H	–	Heel
I	–	Interference or in Gage
J	–	Journal
L	–	Locked or Lost
M	–	Middle Row
N	–	Nose
O	–	Out of Gage
R	–	Rollers

General Nomenclature (Cont.)

S	–	Seal
SP	–	Spearpoint
ST	–	Shirt-Tail
T	–	Teeth or Inserts, Tracking
W	–	Wear

CUTTING STRUCTURE

BT	–	Broken Teeth or Inserts
BU	–	Balled Up
CT	–	Chipped Teeth
LT	–	Lost Teeth or Inserts
TT	–	Tracking Teeth
UW	–	Uniform Wear
WT	–	Worn Teeth

BEARING CONDITIONS

BR	–	Broken Rollers
LB	–	Lost Bearings
LR	–	Lost Rollers
SE	–	Seals Effective
SF	–	Seal Failure
SQ	–	Seals Questionable

CONE SHELL CONDITIONS

CA or AC	–	Cracked Axially
CC	–	Cracked Circumferentially
EC or CE	–	Cone Erosion
CI	–	Cone Interference
LC	–	Lost Cone

HOW TO REPORT BIT WEAR IN EIGHTHS

TOOTH DULLNESS	MILLED TOOTH	INSERT BITS
T1	Tooth height 1/8 gone	1/8 of inserts lost or broken
T2	Tooth height 1/4 gone	1/4 of inserts lost or broken
T3	Tooth height 3/8 gone	3/8 of inserts lost or broken
T4	Tooth height 1/2 gone	1/2 of inserts lost or broken
T5	Tooth height 5/8 gone	5/8 of inserts lost or broken
T6	Tooth height 3/4 gone	3/4 of inserts lost or broken
T7	Tooth height 7/8 gone	7/8 of inserts lost or broken
T8	Tooth height all gone	All of inserts lost or broken

NOTE: If any one row has a majority of teeth broken, add the letters ("BT").

BEARING CONDITION (check the worst cone)

B1	bearing life used:	1/8
B2	bearing life used:	1/4 (tight)
B3	bearing life used:	3/8
B4	bearing life used:	1/2 (medium)
B5	bearing life used:	5/8
B6	bearing life used:	3/4 (loose)
B7	bearing life used:	7/8
B8	bearing life all gone	(locked or lost)

BIT GAGE

I	....	in gage
O	....	out of gage

(NOTE: If out of gage, follow the "O" with the amount of gage in fractions of an inch)

Example #1   T2-B4-I
    (Teeth 1/4 gone, or inserts 1/8 lost or broken, bearing medium, bit in gage)

Example #2   T6BT-B6-0 1/2
    (Teeth 3/4 gone, or inserts 3/4 lost or broken, bearing loose and bit out of gage 1/2 inch).

Figure 5.23. IADC recommendations for grading dull bits

BIT RECORD

DRIVE SIZE & TYPE		O.D.	DRILL COL.	NO.	O.D.	I.D.	PUMPS	LINER	DRAWWORKS & POWER	FUEL	WATER
1.	4½ XH	6¼		13	8 & 10	2¾	1 EMSCO D8700	5½	EMSCO 3 RIO.	N-GAS	HAUL
2.				21	7		2 EMSCO D8700	5½	800-A L 3000		

	AREA	STOCKPOINT NO.	DO NOT USE		STATE	ZONE	COUNTY	CONTRACTOR	OPERATOR	DATE	PURCHASER
M Wilson	West Tex - New Mex	1HN31									

USE PE	USE NO	NO.	SIZE	MAKE	TYPE	JET 32ND IN	SERIAL	DEPTH OUT	FEET	HOURS	FT HR	ACCUM DRLG HRS	WT 1000 LBS	RPM	VERT DEV.	PUMP PRESS	SPM 1	SPM 2	MUD WT.	MUD VIS.	MUD W.L.	DULL. COND. T	B	G	OTHER	FORMATION CALL DATES REMARKS
		1	17½	HTC	OSC3A	3-11	RETIP	425	425	11½	40		AL		¼	1100										
							SET 13⅜" CASING 1-21-75																			
		2	12¼	REED	Y11-J	3-11	226114	2114	1685	40¼	42	6	55	60	¾	1400	60		BRINE			6	8	¼		
		3	12¼	HTC	OSC1G	3-11	SB645	3141	1027	34	30		55	60	1°	1550	60		BRINE			6	8	F		
		4	12¼	HTC	RX44	3-11	RERUN	4124	983	71½	13		45	40	1°	1600	62		BRINE			7	8	F		1/30 4323'
		5	12¼	HTC	OWV-J	3-11	HS094	4350	226	20¼	113		55	62		1400	60		BRINE			5	5	F	SF1 STW	
							SET 9⅝" CASING 1-31-75																			
		6	8¾	HTC	J55	3-11	HE645	7516	3166	179¼			15/40	50	3°	1700	50		BRINE			3	1	F	SE123 U-3	2/6 6695'
	/	7	8¾	HTC	J55	3-11	HE642	9187	1671	222			24/40	40	1½	1850	58		9^6	40	10	3	3	1/16	SE123 BT U-3	2/17 8792'
	/	8	8¾	HTC	J33	3-11	AX996	10,035	848	103			40	50	3½	1900	46		9^6	40	10	6	4	F	SE123 U-3	2/21 9338'
		9	8¾	HTC	J77	3-11	HH907	10,065	30	28½			40	50		1850	58		9^6	54	10	1	1	F	SE123 U-3	2/24 9961'
		10	8¾	HTC	J55	3-11	HE659	10,177	142	26			40	40	3¼	1800	56		9^6	54	10	8	8	½	RIG	2/27 10,017'
	/	11	8¾	HTC	J77	3-11	RERUN *9	10,745	568	119½			45	40		1800	56		9^6	45	8	6	8	F	SQ12 STW U-1	3/3 10,560'
		12	8¾	HTC	J77	3-11	FD782	11,000	255	52½			45	40	1½	1800	56		9^6	40	8	2	2	1/16	SE123 U-3	3/7 10,976'
	/	13	8¾	HTC	J55	3-11	BM627	11,812	812	124			40	40	1¾	1800	56		9^6	40	10	8	8	¾	DST #1 10,950 to 11,000	3/11 11,408'
	/	14	8¾	HTC	J55	3-11	RERUN	11,913	101	49			40	45		1800	56		9^6	40	10	8	8	1/16		3/14 11,698
		15	8¾	REED	FP64	3-12	RERUN	12,030	117	29¼			40	45		1800	60		9^4	40	8	5	8	F	SF13 U-1	
		16	8¾	HTC	J55	3-12	RERUN *6	12,110	80	10¼			35	40		1800	60		9^6	49	9^8	3	5	F	SE123 U-3	3/20 12,041'

BIT CONDITION CODE: RP – REPAIRED RR – RERUN

Figure 5.24. Bit record of a West Texas well

Bit no. 7 on the bit record (fig. 5.24) is a Hughes J55. It drilled 1,671 feet in 222 hours. Assuming the information given previously, this hole cost $29.86 per foot. In comparison, bit no. 8, a Hughes J33, made 848 feet in 103 hours for a calculated cost of $30.78 per foot, according to the information given.

Care and Maintenance

Although a rock bit is about as foolproof as any piece of equipment on the rig, it can be abused to such an extent that its useful life is cut short. Certain general procedures can be helpful in prolonging its life.

Figure 5.25. Bit breaker

Pointers for Making Up the Bit

1. If the box containing the bits has been opened previously, check the threads on the bit shank, and clean them if necessary.
2. Dope the threads with a good-quality, clean thread lubricant of the type recommended for tool joint threads.
3. Be sure to use the correct breaker plate for the size and type of bit being run (fig. 5.25).
4. Cover the hole, and place the bit breaker in the locked rotary table. Small bits may be started by hand-screwing them into the box threads on the collar sub. Place the bit in the breaker, and lower the collar over the shank. Rotate the collar by hand until it shoulders onto the bit.
5. Place the makeup tongs on the collar just above the bit. Make up to the proper torque. Never use a sledgehammer on a bit or on the tongs in an effort to tighten the bit.

When *running the bit to bottom,* if there are ledges or boulders sticking out into the hole, work the bit past them. If the hole is undergauge, especially near bottom, ream down to bottom with due care. If there is likely to be a fill of cuttings settled on bottom while the pipe is out of the hole, run the bit in carefully, with full circulation to clean out. Jamming the bit into the fill poses the threat of balling up the bit and jamming the cones until they start dragging.

Changing nozzles and *reusing insert bits* are often possible. Nozzles used in jet bits are made of extremely hard material to prevent erosion. They are therefore very brittle and may be chipped while being changed on the rig. Study and follow the manufacturer's recommendations on changing procedures for the particular bit-design nozzles.

Insert bits equipped with journal bearings can be rerun if the bearing seals have not been damaged and if the inserts are almost all present and in good condition. They are not routinely returned to the manufacturer for repair if they are considered suitable for rerun. Examine the seals for leakage or damage, and the cones for cracks and erosion. If the grease diaphragms are in good condition, fill the reservoirs with lubricant; or, if the reservoirs still contain grease, rerun the bit. Do not store a sealed bearing bit in diesel fuel or any other type of oil. If necessary, store the bit in fresh water.

DIAMOND BITS

Economics

Compared to other bits, a diamond bit is quite expensive. It may cost three or four times as much as a carbide insert bit, just as the latter may cost several times as much as a steel-tooth bit. But the diamond bit, like the various rock bits, is used when it can be shown to offer an economic advantage over others.

When making comparisons among the many bits in regard to cost per foot of making hole, the following factors have to be included in the calculations:
(1) the net cost of each bit considered;
(2) the cost of one or more round trips in dollars for rig time;
(3) the cost of rotating time calculated in dollars per hour; and
(4) the added cost of mud when making trips.

All the factors are added together, and the sum is divided by the number of feet drilled by each bit being considered.

As a general rule of thumb, one considers using a diamond bit¹ when a rock bit's production rate falls below 10 feet per hour, or² when the hole diameter is less than 6 inches. Using a diamond bit offers the advantages of remaining in the hole during high-pressure or possible blowout conditions, being able to reduce mud weight or to condition mud without losing circulation, avoiding trips during bad weather, and reducing rig wear. *Note*

In addition, the diamond bit has a salvage value in the form of serviceable diamonds left in the bit when it is pulled. This salvage value, when deducted from the original cost, gives the net cost of the bit. Even so, a diamond bit is generally more expensive than a rock bit. The most important factor in its favor is the fact that it makes more hole than any other bit over the entire period of its rotating life. Although the diamond bit lasts longer and requires fewer round trips for bit change, it must still make hole at a reasonable rate; otherwise, time lost in rotating will cancel out the savings in round trips for bit change.

Design

Diamond bits have very basic designs and no moving parts, both desirable characteristics.

Watercourses in diamond bits are not variable as they are in jet nozzle bits. They have two basic configurations – cross-pad flow and radial flow. There are also variations of each, as well as combinations of both.

Figure 5.26. Soft-formation diamond bit (*Courtesy of NL Hycalog*)

Figure 5.27. Diamond bit for soft to medium-hard formations (*Courtesy of Christensen Diamond Products*)

Figure 5.28. Hard-formation diamond bit (*Courtesy of NL Hycalog*)

Diamond bit *cones* come in a variety of configurations. The three most popular are the double cone, the modified double cone, and the round cone.

The *double cone* is used chiefly for soft, medium-soft, and medium formations such as shale. This type of bit has the sharp, pointed nose characteristic of bits designed for softer formations. The *modified double cone* has a broader nose and is used for harder formations. The *round cone* configuration is used for hard to very hard formations; it usually has a very slow penetration rate but allows a large amount of diamond coverage in the nose area of the bit.

The size and spacing of diamonds on a diamond bit determine its use. Diamonds are classified by number of carats, a measure of weight; the larger the diamond, the more carats it weighs.

Very large, widely spaced diamonds are used for cutting large pieces of soft sand and shale. The diamond size is usually 2 to 5 carats (fig. 5.26).

Medium spacing of large ¼- to 1-carat diamonds is used in bits for more versatile use (fig. 5.27). This type of bit can be used for a wide range of sand, shale, and limestone formations.

Hard formations call for smaller diamonds — ⅛ to ½ carat — set in closer patterns (fig. 5.28). The denser diamond placement produces longer wear in sharp, hard formations such as dolomites and hard shales.

Diamond bits have three main cutting actions. They are *compressive* (where the formation is cracked), *abrasive* (where the formation is worn away), and *plowing* (where the formation is ruptured).

Use in Drilling

The following notes about drilling practices using diamond bits have been excerpted from the *IADC Drilling Manual*.

The IADC stresses the importance of *maintaining a clean hole*, no matter what type of bit is being used. But keeping the hole clean is particularly important when using a diamond bit, in order to prevent damage to such an expensive item. Diamonds are especially vulnerable to loose junk iron, so every precaution should be taken to clean the hole thoroughly before diamond drilling is started and to keep it clean during the operation. Normally, it is not difficult to keep a hole clean during rock bit drilling. Following the suggestions listed will eliminate the cost of extra trips for cleaning.

(1) If rock bit cones are lost, fish out cones and bearings at that time. Otherwise, bearings become embedded in the walls or stored in cavities and may fall to the bottom during a trip, damaging the diamond bit.
(2) Keep tong dies securely keyed in place.
(3) Keep a wiper on the pipe when going into or coming out of the hole.

(4) Use a junk sub or similar tool with the last two or three rock bits. This procedure has proven effective in ensuring a clean hole ready for a diamond bit. If there is doubt about whether or not the hole is clean, run a magnetic fishing tool to recover loose iron particlees in the hole.

Balancing Mechanical and Hydraulic Factors

When going to bottom with a diamond drill bit, the bit should be stopped at least 2 feet off bottom and the pump speed should be regulated to deliver the necessary volume for the bit selected. The bit should be lowered to the bottom without being rotated, if possible, in order to pump any junk iron or pieces of formation off the bottom, applying from 5,000 to 8,000 pounds of weight to ensure that the bit is set on bottom and not in cavings. After this operation, the bit should be raised again to about 2 or 3 feet from the bottom, started slowly (40–50 rpm), and then again lowered to bottom as weight is applied (about 5,000 pounds). The drilling weight should then be gradually increased until the best penetration rate is obtained. The pressure established after the bit has started drilling should be kept in mind constantly while the bit is on bottom. If the final pump pressure increases or decreases, there is definite indication that something abnormal is occurring, and the cause should be determined and corrected. Otherwise, the diamond bit may be damaged, or costly rig time may be needlessly used.

Wear and Problems

Diamond bits are expensive pieces of equipment; care should be taken to get maximum use from them. Attention to certain precautions in their use can prevent much damage and wear.

Drilling junk at the bottom of the hole shears diamonds at the surface of the bit matrix and eventually grooves the matrix itself. The hole should be perfectly clean before running a diamond bit.

Drilling with no fluid produces intense heat that turns cuttings and bit matrix into a molten state. Diamonds become dislodged and travel around in the hole, destroying the bit completely. Although such an occurrence is rare, its variations are common. Inadequate fluid hyraulics for cooling and cleaning the bit can result in partial or total plugging of its fluid passageways, causing cuttings to ball up and clog the bit. The heat generated causes the cuttings to burn into the matrix and dislodge diamonds; it can also crack the matrix.

A combination of heat, too little drilling fluid, and junk in the hole can cause heat cracks and checks in the matrix. If drilling continues after diamond damage has occurred, the bit can be a total loss.

Diamond bits are strong and will serve well if they are properly cared for. The most important things to watch for when drilling are using correct hydraulics, selecting the right bit for

Figure 5.29. Original fishtail bit

Figure 5.30. Jet modified from a fishtail bit

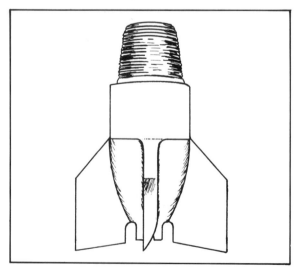

Figure 5.31. Four-way drag bit

the formation being drilled, keeping a clean hole, and keeping an eye on the rate of penetration so that any disruption is quickly noticed and steps can be taken to correct any developing problems.

DRAG BITS

Although the grandfather of all rotary drag bits—the fishtail—has been relegated to the museum of petroleum antiquities, not all drag bits have met the same fate. There are a few areas left where the drag bit has a recognized place in the scheme of things.

The original type of fishtail bit had two blades and an eye for the drilling mud near the threaded shank of the bit (fig. 5.29). This bit ruled the oil fields from the time of Spindletop until the 1920s. During its heyday, every toolpusher was a bit designer and every blacksmith a bit manufacturer. Few drillers were so lacking in imagination that they could not come up with an idea to transform the common fishtail bit into a real hole-maker.

The first man to modify the fishtail by bringing the watercourses through the blade of the bit to an outlet (fig. 5.30) was the inventor of the jet bit prototype. The first actual jet bit, produced in 1947, was also a fishtail drag bit, but probably resembled the one shown in figure 5.31, with three or four blades of hard metal welded to the body of the bit. Bits of this type are occasionally used today for drilling soft, shallow formations before setting surface casing. They are commonly available in sizes up to 24 inches.

Replaceable-blade bits are made in sizes from 1⅞ to 16 inches for shot-hole, exploratory, slim-hole, blasthole, and water-well drilling. Various sizes of bit bodies can be obtained for the complete range of blade sizes. Blades are expendable and are designed to be completely worn out and thrown away.

LESSON 5 QUESTIONS

Put the correct answer in the blank before each question. If there is more than one correct answer, put in all the correct letters. If a blank is drawn in the question, write out the answer as well as supply the letter in the multiple choice slot. The very act of writing down the answer will help you remember it.

_____ 1. Which of the following is the cutting or boring element used in drilling?
    A. Mud
    B. Bit
    C. Journal bearings
    D. Inserts

_____ 2. Openings or passageways in bits that permit drilling fluid to enter the wellbore under pressure are –
    A. jet nozzles
    B. junk slots.
    C. inserts.
    D. intermeshes.
    E. none of the above.

_____ 3. The measure of a bit that determines the size of hole it drills refers to the bit's –
    A. ROP.
    B. bradding.
    C. gauge.
    D. type.
    E. interfit.

_____ 4. If the weight on a tooth in a bit has been substantial enough to cause the tooth to become dull and the softer inner portion of the tooth to cover over the harder case area, the condition is referred to as –
    A. inner bearing failure.
    B. unbalanced tooth wear.
    C. tracking.
    D. bradding.
    E. tooth abrasion.

_____ 5. Two-cone bits are used in –
    A. large-diameter holes.
    B. directional drilling.
    C. both of the above.
    D. neither of the above.

_____ 6. The gauge surface of a bit determines –
   A. the type of formation the bit may be used to drill.
   B. rate of penetration.
   C. wear index.
   D. its resistance to spalling.
   E. the size of the hole drilled.

_____ 7. Rate of penetration (ROP) refers to –
   A. how fast the bit is making hole.
   B. how fast the bit is rotated, in rpm.
   C. A and B above.
   D. none of the above.

_____ 8. In most situations, the driller tries to drill as many feet as possible with one bit.
   A. True
   B. False

_____ 9. Roller cone bits, or rock bits, are generally classified as _____

   _____ and _____.
   A. diamond bits and milled bits
   B. drag bits and steel-tooth bits
   C. jet-deflection bits and diamond bits
   D. milled bits and tungsten carbide bits
   E. tungsten carbide bits and insert bits

_____ 10. Tungsten carbide insert bits are rapidly replacing hard-formation steel-tooth bits.
   A. True
   B. False

_____ 11. Which of the following factors affect a bit's performance?
   A. The type of formation drilled
   B. The rate of penetration
   C. The weight on bit
   D. All of the above

_____ 12. To be an effective cutting tool in rotary drilling, the bit must –
   A. be new.
   B. be changed each time a trip is made.
   C. have diamonds embedded in its bottom and sides.
   D. be rotated.
   E. be none of the above.

_____ 13. The characteristics of diamond bits that distinguish them from roller cone bits are that—
   A. tungsten carbide buttons are embedded in their cones.
   B. diamond bits have no cones or teeth.
   C. diamond bits have four cones.
   D. diamonds are embedded in the bottom and sides of diamond bits.
   E. all of the above are true.

_____ 14. A feature common to all bits is their—
   A. capacity to drill all formations.
   B. cutting and gouging action during drilling.
   C. cones and teeth.
   D. passageways through which drilling fluid flows.
   E. having all of the above features.

_____ 15. Rock bits were used to make hole over fifty years ago but they did not have—
   A. two cones.
   B. interfit.
   C. steel teeth.
   D. jet-bit arrangement watercourses.

_____ 16. A bit with long, widely spaced teeth might best be used in drilling—
   A. hard formations.
   B. medium formations.
   C. soft formations.
   D. all of the above.

_____ 17. Tungsten carbide insert bits have been improved in recent years. All of the following are improvements made to the bit except—
   A. development of sealed bearings.
   B. wear-resistant materials in cones.
   C. jets providing more safety for inserts.
   D. need for less weight on bit (WOB).
   E. ability to run at high rotary speeds.

_____ 18. In addition to roller bearings, the two other major classes of bearings used in bits

   are _____ and _____ bearings.
   A. sealed and journal
   B. self-lubricating and ring
   C. nose and ring
   D. friction and ball

_____ 19. Some bearings are self-lubricating.
   A. True
   B. False

_____ 20. Drilling fluid provides lubrication for bearings in some bits.
     A. True
     B. False

_____ 21. Sealed bearings utilize a *lubricant reservoir* to lubricate bearings.
     A. True
     B. False

_____ 22. Sealed roller bearings were used first in steel-tooth bits.
     A. True
     B. False

_____ 23. Rock bits that have journal bearings use a specially made grease to resist galling.
     A. True
     B. False

_____ 24. Journal bearings tend to spall.
     A. True
     B. False

_____ 25. When drilling brittle rock, if the WOB in a rock bit is increased by 35 percent, the rate of penetration would probably –
     A. remain the same.
     B. increase by one-third.
     C. double.
     D. decrease by half the percent increase of the WOB.

_____ 26. In order to allow for fast penetration in soft nonabrasive formations, a driller using a steel-tooth bit with reduced WOB might –
     A. replace the steel-tooth bit with a tungsten carbide bit.
     B. increase rotary speed.
     C. decrease rotary speed.
     D. increase the WOB.
     E. replace the steel-tooth bit with a diamond bit.

_____ 27. Features of the jet-bit nozzle that distinguish it from watercourses in a conventional bit are that the –
     A. fluid hits cutting surfaces to clean them.
     B. fluid creates turbulence around the bit.
     C. fluid velocities are high.
     D. fluid lifts cuttings up the annulus.

_____ 28. Directional drilling in soft formation is sometimes done using _____.
     A. fishtail bits
     B. jet air bits
     C. jet-deflection bits
     D. two-cone bits

_____ 29. Bearing failure or wear is likely to be caused by –
   A. reaming an undergauge hole.
   B. drilling a new type of formation too hard for the bit.
   C. excessive rotation time coupled with heavy weight.
   D. abrasive materials in the drilling fluid.
   E. none of the above.

_____ 30. If _balling-up_ occurs during the drilling operation, the driller can remedy the situation by –
   A. increasing the WOB.
   B. decreasing rotary speed.
   C. decreasing the WOB.
   D. increasing volume and/or velocity of drilling fluid.
   E. replacing the bit with a new one.

_____ 31. A bit was graded T4-B2-I. The condition of the bit is best described as "teeth one-half gone (or inserts one-half lost or broken); bearings tight; and the bit in-gauge."
   A. True
   B. False

_____ 32. If a bit is graded T6-B5-0½, the statement that best describes the condition of the bit is –
   A. teeth three-fourths gone (or inserts three-fourths lost or broken); bearings medium; bit out-of-gauge one-half inch.
   B. teeth three-fourths gone (or inserts three-fourths lost or broken); three-eighths of bearing life remaining; bit out-of-gauge one-half inch.
   C. teeth or inserts three-fourths gone; bearings loose; bit in-gauge.
   D. teeth one-fourth gone (or inserts three-fourths lost or broken); bearings loose; bit out-of-gauge one-half inch.

_____ 33. Bit no. 10 on the bit record (fig. 5.24) was run for twenty-six hours (26 hrs.). The average rate of penetration of this bit was approximately –
   A. 113 feet per hour.
   B. 15 feet per hour.
   C. 11 feet per hour.
   D. 110 feet per hour.
   E. 5 feet per hour.

_____ 34. When bit no. 10 on the bit record (fig. 5.24) was pulled and graded, the condition of the bit's teeth and bearings was –
   A. teeth half gone, or inserts five-eighths lost or broken.
   B. teeth five-eighths gone, or inserts five-eighths lost or broken.
   C. teeth five-eighths gone, or inserts five-eighths lost or broken; three-fourths of bearings' life remaining.
   D. teeth all gone, or inserts all lost or broken; bearings' life gone.

_____ 35. The WOB of bit no. 10 was –
    A. 40,000 lbs.
    B. 35,000 lbs.
    C. 62,000 lbs.
    D. 55,000 lbs.
    E. none of the above.

_____ 36. Bit no. 10 drilled 142 feet in 26 hours. Assuming that the following information is true – rig operating costs (OC) = $201/hr.; bit cost (B) = $1,000; round trip time (T) = 10 hours – the cost per foot of hole drilled using bit no. 10 was –
    A. $30 per foot.
    B. $58 per foot.
    C. $35 per foot.
    D. $47 per foot.
    E. $50 per foot.

_____ 37. The rotary speed of bit no. 10 during its use was –
    A. 62 rpm.
    B. 50 rpm.
    C. 40 rpm.
    D. 45 rpm.
    E. 60 rpm.

_____ 38. All of the following are helpful pointers for making up the bit except:
    A. Dope threads with a good quality lubricant.
    B. Use a sledgehammer to assist in tightening the bit.
    C. Cover the hole and place the bit breaker in the locked rotary table.
    D. Place makeup tongs directly on the bit.

_____ 39. Insert bits can not be rerun.
    A. True
    B. False

_____ 40. Diamond bits may cost four times as much as tungsten carbide insert bits.
    A. True
    B. False

_____ 41. Each of the following is an advantage of using a diamond bit except:
    A. It reduces rig wear.
    B. It can remain in the hole during high pressure or blowout conditions.
    C. It conditions mud without losing circulation.
    D. It requires fewer round trips.
    E. It costs more to purchase than a roller cone bit.

_____ 42. Watercourses in diamond bits have two basic configurations. They are called

_____ and _____.
   A. radial and cross-pad flow
   B. modified and rounded flow
   C. double and cross-pad flow
   D. center and off-centerline flow

_____ 43. The three main cutting actions of diamond bits are abrasive, plowing, and _____

_____ action.
   A. gouging
   B. biting
   C. compressive
   D. breaking
   E. dislodging

_____ 44. If drilling with a diamond bit is allowed to progress with inadequate fluid hydraulics for cooling the bit, which of the following is most likely to occur?
   A. Total plugging of fluid passageways
   B. Balling-up
   C. Heat build-up
   D. Dislodging of diamonds
   E. All of the above

_____ 45. Drag bits are occasionally used today to drill –
   A. hard formations.
   B. soft shallow formations.
   C. formations having abnormal pressure.
   D. all of the above.

_____ 46. In order to prevent damage to a diamond bit, several precautions should be taken. The most important of these is probably –
   A. drilling only in soft formations.
   B. keeping the hole clean.
   C. reaming a hole cut by a rock bit.
   D. all of the above.

_____ 47. In drilling a soft formation using a diamond bit, the bit used would most likely be –
   A. double cone.
   B. modified double cone.
   C. round cone.
   D. triple cone.
   E. none of the above.

_____ 48. The size and spacing of diamonds on a diamond bit determines its use. Very large and widely spaced diamonds would probably be used to drill –
   A. limestone formations.
   B. hard formations such as dolomites.
   C. fine hard sands.
   D. soft sands and shale.

_____ 49. When the hole is drilled at a speed that is not fast enough to hold the bit, the resulting problem would most likely be –
   A. center-coring.
   B. off-center wear.
   C. cone erosion.
   D. insert breakage.

_____ 50. Center-coring is a wearing away of the nose area of a cone. Excessive weight applied to the nose area of the cone will result in –
   A. cone-skidding.
   B. cone breakage.
   C. loss of inserts.
   D. balling-up.